BOTANY FOR ALL AGES

"A super book! A must for all teachers in helping students discover the wonders of nature. There is something for everyone in this book."

—Nancy D'Arcy, teacher, St. Louis Public Schools and Missouri Botanical Gardens

"Jorie's book is filled with a wealth of useful tidbits of information about the plant world and ways of communicating that information to children, indeed in actively involving children in investigating the 'strange, green aliens' among us. Although the botanical material in the book will be of interest to people of any age, Jorie Hunken's insights into ways of introducing the plant world to children make this a special book for teachers and youth leaders."

—Charles E. Roth, Director of Educational Services, Massachusetts Audubon Society and author of *Plant Observers Guidebook*

BOTANY FOR ALL AGES

*Learning about nature
through activities using plants*

Jorie Hunken

and

The New England Wild Flower Society

The Globe Pequot Press

Chester, Connecticut

Library of Congress Cataloging-in-Publication Data

Hunken, Jorie.
    Botany for all ages: nature activities in the world of plants/by Jorie Hunken and the New England Wildflower Society.—1st ed.
      p.   cm.
    Includes index.
    ISBN 0-87106-616-5
    1. Botany—Study and teaching—Activity programs. 2. Botany—Experiments. 3. Envrionmental education—Activity programs.
I. New England Wildflower Society. II. Title.
QK52.55.H87 1989
581'.07'8—dc19

                                                    88-37083
                                                      CIP

Photographs on pages v, 15, and 28 reprinted courtesy Peter Southwick. Photograph on page 81 reprinted courtesy Jean C. Tibbils. Other photographs are by the author.

Manufactured in the United States of America
First Edition/Second Printing

In memory of my friend Jeanie Tibbils, whose photographic skills add so much to this book and whose artistry, humor, and free-flowering creativity added so much to the lives of her many friends and students

Contents

Foreword by Thomas Buchter xiii

Acknowledgments xv

Introduction xvii

Introduction to Teaching With Plants 1

1 - *Becoming Environmentally Aware through Activities Using Plants* 3

2 - *Creating an Environment for Learning* 4

3 - *When the Learners Teach the Teachers* 6

4 - *Botany for All: Special-Needs Children Out-of-Doors* 7

5 - *Structuring the Excursion* 8

6 - *Up Your Sleeve: Devices to Enhance Learning Situations* 10

7 - *Individual Learning Styles: Different People Learn in Different Ways* 10

8 - *The Art of Learning* 11

9 - *Ideas for Art Projects Using Plant Material* 12

10 - *An Example of an Art Activity: Dyeing with Plant Materials* 15

11 - *Activities for the Senses and the Imagination* 18

12 - *Inventing Seeds That Can Travel* 20

13 - *Ways to Exercise the Senses* 21

14 - *Enhancing Observational Skills* 22

15 - *Centering* 24

16 - *Approaches to Centering* 24

17 - *The Learner as Teacher and Vice Versa* 25

18 - *To Have and to Hold* 26

19 - *Scavenger Hunts: Creative Collecting* 27

20 - *Developing Environmental Values* 28

21 - *Do You Have to Know the Names? A Rationale for Correct Identification* 29

22 - *Activities for Identifying Species* 30

23 - *Developing a Vocabulary for Appreciating and Identifying Plants* 31

24 - *Creating a Key for Species Identification* 32

Activities for Learning About Plants 35

25 - *Seaweed Prints* 37

26 - *Fern Activities* 38

27 - *Mushroom Collecting* 39

28 - *Haircap Moss Rain Meter* 40

29 - *To Pick or Not to Pick a Flower* 41

30 - *Tiny Flowers on the Playground* 42

31 - *Learning the Names for Parts of Flowers* 43

32 - *Sorting Flowers* 45

33 - *Comparing Garden Flowers with Wild Relations* 46

34 - *Inventing Flowers* 46

35 - *Calendar of Bloom* 47

36 - *Nectar Guides—Road Signs for Pollinators* 47

37 - *Trick-a-Bee: Designing Flowers to Attract Bees* 49

38 - *Being Bees—Exploring Pollinating Flowers* 50

39 - *Weather Conditions and Pollinating Insects* 51

40 - *Violets—Discovering the Hidden, Self-Pollinating Flowers* 52

41 - *Wind-Pollinated Flowers* 53

42 - *Grass Watching* 54

43 - *The Giant Female Flower: Corn* 55

44 - *Connecting Flowers with Seed Production* 56

45 - *Self-Pollinating an Amaryllis to Produce Fertile Seeds* 57

46 - *The Kitchen as a Source of Seeds* 58

47 - *Activities Using Collections of Seeds* 59

48 - *Growing Pot Herbs from Seeds* 60

49 - *Which Are Fruits, Which Are Vegetables?* 60

50 - *Hidden Seeds in the Winter Earth* 61

51 - *Growing Peanuts from Seeds* 62

52 - *Is It Ripe?* 63

53 - *Windborne Seeds* 64

54 - *Seeds That Hitchhike* 65

55 - *Seeds That Float in Water* 66

56 - *Floating Cranberries* 67

57 - *Berries and the Beasts* 67

58 - *Plants and Their Offspring—A Search for Origins* 68

59 - *Sumac-Ade* 69

60 - *Acorn Play* 70

61 - *Pinecone Rain Meter* 72

62 - *Making Raisins* 72

63 - *Pumpkins: Beyond the Jack-O'-Lantern* 73

64 - *Ripening Fruit Using Ethylene Gas* 74

65 - *A Coat for a Seed* 74

66 - *Sprouts: Meeting Them and Eating Them* 75

67 - *Sprouting Gestures* 77

68 - *Cooked Beans/Live Beans* 78

69 - *Experiments in Finding Starch in Plants* 78

Various Parts of Plants: Names and Functions 81

70 - *Water Movement Through Plants: The Veins* 83

71 - *Making Leaf Skeletons* 84

72 - *How the Leaves Help Pull Water Through the Plant* 85

73 - *Gases Also Come Out of Leaves* 86

74 - *A Leaf Is Full of Water* 87

75 - *A Wilting Contest* 88

76 - *Using Leaves in Art Projects* 88

77 - *Response to Stem Damage* 91

78 - *Does a Plant Need to Breathe?* 92

79 - *The Holding Power of Roots* 93

80 - *Carrot Machine* 94

81 - *Responses in Roots* 95

82 - *Some Food Crops That Also Serve as Storage Areas for Plants* 97

83 - *A Rose is a Rose is a Cabbage Bud* 98

84 - *Paperwhites: Growing Spring Flowers from Bulbs Indoors* 99

85 - *Using Information on Plant Parts in a Game* 100

86 - *Bring Trees into the Classroom* 100

87 - *How Many Leaves on a Tree?* 101

88 - *The Autumn Leaves* 102

89 - *Estimating the Height of a Tree* 103

90 - *Looking into Tree Buds* 104

91 - *Twigs* 105

92 - *Finding the Ages of Young Pine Trees* 106

93 - *Bark Cork* 107

94 - *Hardwood/Soft Wood* 108

95 - *Explorations Using Lumber and Cross-Sections of Wood* 109

96 - *Comparing Leaves of Deciduous Trees and Evergreens* 110

97 - *General Instructions for Growing Plants in Pots* 111

98 - *Experiments with Growing Plants* 114

99 - *Propagation of Plants from Cuttings: New Plants from Old* 116

100 - *Poking into the Soil* 117

101 - *Soil As an Art Medium* 120

102 - *Acid Soil or Alkaline Soil* 121

103 - *Plant Food: Fertilizer Supplements* 122

104 - *Open Terrariums* 123

105 - *The Effects of Sunlight or Soil on a Plant Species* 124

106 - *Body Language in Plants* 126

107 - *Plants That Move on Their Own* 126

108 - *Row Plantings as Environmental Indicators* 127

109 - *Plants as Indicators of Roadside Pollution* 128

110 - *Charts and Graphs: Facts at a Glance* 129

111 - *Plant Math: Growth Curves and Averaging Data* 130

112 - *Activities to Practice the Art of Careful Looking:*
     Transects and Quadrats 131

113 - *Background Activities for Understanding Plant Succession* 132

114 - *Developing an Awareness for a Habitat:*
     Ways of Describing and Giving It Value 133

115 - *Seasonal Changes in a Special Plant* 133

116 - *Dandelions from Top to Bottom* 134

117 - *Winter Rosettes: Wild and Cultivated* 136

118 - *Outdoors in the Rain* 137

119 - *Plants as Historical Indicators* 138

120 - *Creating Imaginary Plants to Fit Special Environments* 138

121 - *Creating Animals to Go with the Plants* 139

122 - *Insects on Plants* 139

123 - *Investigating a Rotting Log* 140

124 - *Beetle Writing Under Bark* 141

125 - *Bird Nests* 142

126 - *Who Eats Whom (with a Plug for Endangered Species)* 143

Appendix 145

Glossary 149

Bibliography 151

Index 155

Foreword

The New England Wild Flower Society is an organization that grew out of a movement to protect native flora that was rapidly being depleted by careless agricultural practices and a florist industry that was removing large quantities of plants for decorative purposes. In 1932, the society was incorporated to promote and encourage the conservation and horticultural uses of native plants through education and research. In the 1930s schools had programs to teach children about wildflowers. The society published leaflets to supplement the programs, emphasizing the need to leave the plants and flowers where they were found so that others might enjoy them and so that the plants would flourish and increase.

At the same time Will C. Curtis, a landscape designer, had purchased a tract of land in Framingham, Massachusetts. Here he planned to establish a wildflower garden to demonstrate to people that native plants were suitable garden subjects and, at the same time, to convey to people that native plants were a precious resource to protect and cherish. Along with his partner, Dick Stiles, they created Garden in the Woods, a garden that achieved national recognition. In the early 1960s concern for the future of the garden became a topic of many discussions.

Through the work of a number of individuals, it was decided that the New England Wild Flower Society, which was based in Boston, would take over ownership of the Garden in the Woods. An endowment was raised to support the garden, and it was opened to the public. Soon after, the society put up a building and moved its headquarters to Framingham. The new executive director and board of trustees recognized that, not only would a staff be needed to maintain and develop the garden, but also staff would be needed to develop an education program that served to promote the society's purposes for existing.

The need to educate children in the stewardship of our natural resources became top priority for the organization and was carried out by Jorie Hunken and the education committee of the New England Wild Flower Society. This book is a result of this unique children's education program. The activities in this book, all based on sensory experiences, will bring the learner into contact with plants to increase an individual's capacity to see and feel at home in the natural world.

Botany for All Ages sets the tone for a common-sense approach to the study of plants. It reaches beyond the usual textbook experiences and provides the opportunity for the child to develop an intimate relationship with plants.

Thomas Buchter, former executive director
New England Wild Flower Society, Inc.

Acknowledgments

Much of this book is based on the ideas generated by a group of people who worked together throughout the 1970s to create an environmental education program based on native plants. Their headquarters was the oak woodland of the Garden in the Woods, Framingham, Massachusetts, the home of the New England Wild Flower Society. It was here that the teachers taught children's classes, took school groups on tours, and wrote the curriculum reports that form the basis and direction of this book. Much of their caring and many of their actual phrases are in these pages. Especially to be thanked is Biz Paynter, who has been waiting for me to write it all down for twenty years.

The book is also an expression of the Wild Flower Society's continued involvement in children's education. Individual members contributed funds that made possible the grant that paid for the research and writing of this text. Other environmental organizations have contributed in different ways: the children's program of Habitat, Inc., in Belmont, Massachusetts, lent pictures of their classes; other photos were taken of a home schooling class in Cambridge, Massachusetts; and the volunteer guides here at the Garden in the Woods and at Arnold Arboretum, Jamaica Plain, Massachusetts, continually inspire me with their enthusiasm to teach.

Introduction

This book is for all who teach or want to learn about our environment. It is specifically for teachers and parents who wish to show children that the natural world is both responsive and vulnerable, that the plants around them have names and histories, are connected to other plants and animals, and have lives that can affect people intimately. For teachers responsible for the formal education of our children, the activities can be used as a basis for integrating the subjects of science, art, writing, language development, social studies, mathematics, and history. For any person who works with groups of children (a camp counselor, a teacher in a preschool or an after-school program), this book is a guide and lesson plan for experience-based activities that encourage individual expression as well as group interaction. The activities are not dependent on language skill for their successful completion and are appropriate for mixed-age groups. For parents, this book offers learning experiences for the whole family on weekends and vacations.

The book's goal is to describe situations that stimulate questions and experiments. It is hoped that children will then want to talk about, write about, and draw the things that they experience.

The activities cover a range of botanical subjects, including the parts of plants (flowers, leaves, seeds, and roots), general horticultural requirements (experiments using soil, light, and moisture), and plants as they relate to humans, animals, and their general environment.

The first part of the book is intended as an introduction to environmental education. It contains short chapters to help any adult become comfortable in taking a child or a whole group out-of-doors. These chapters include examples of approaches that have been successful in introducing specific aspects of plant study, hints on structuring a class excursion, a list of handy teaching tools, and a number of short activities to enhance sensory perceptions. There are also several chapters that introduce the concepts of individual learning styles, art activities for stimulating observation skills, and ways of turning a learner into a teacher. The final chapters of the first part deal with the controversies of collecting by picking and the values of learning the names of plants.

The main body of the book consists of specific activities, grouped under six general concepts. Activities for learning the names and functions of the parts of plants are first, followed by chapters that deal with growth requirements, seasonal changes, and interactions between the plant and its environment. Interspersed are uses of plants. It is recommended that you make use of the index in the back of the book to cross-reference topics of special interest.

In a number of the activities, specific age groups are suggested for their use, but there is no listing according to age for each activity. Some young children have sufficient experiential bases to proceed into involved experiments. Some older learners may feel awed by scientific terminology and prefer to do more hands-on activities before venturing farther. Learning about plants should always include a variety of sensory experiences: watching plants grow under different conditions, witnessing the effects of the seasons on particular plants, and observing the relation of plants to different environments. It is assumed that the teacher or adult in charge is the best judge of which activities are appropriate. To help a teacher who must plan a curriculum based on grade level, a general guide to the appropriate kinds of activities for general age groups is included in the appendix.

Stating objectives beforehand gives a means of evaluating the activity's effect on an individual and on a group. The objectives can range from "want the kids to have a good time and pull together as a group" to "help them develop attitudes of stewardship and a vocabulary for describing their feelings and responsibilities." Don't, however, let an objective get in the way of what a child needs.

As a teacher, your actions will always affect the children. Try to keep track of what the kids are *really* learning. Ask the children to describe their experiences at the end of the activity and listen carefully to the answers. (Example: "What was one thing you did or saw on your walk?") Many of the chapters include a few sample discussion questions listed at the end. They are not for testing the children so much as they are opportunities for participants to share their experiences and practice verbalizing concepts and feelings.

It is hoped that everyone who uses this book is able to become a learner, is encouraged to put aside the onerous responsibility of being "the one with all the answers," to become a wonderer, a questioner, a creator of insights, and an opener of doors. Once children sense that you will *help* them learn instead of just teaching them, then you all will learn.

Introduction to Teaching with Plants

1

Becoming Environmentally Aware Through Activities Using Plants

A plant in brightest blossom can engage our attention as surely as any work of art. If a flower is especially beautiful, we want to know how it got there and what its name is. If it is familiar, we recall other places we have seen it and perhaps how we felt at the time. We enjoy learning the derivations of its name, especially when the name reveals uses or ideas from cultures other than our own.

We all live where plants grow. Each has a name and therefore a history of associating with humans. For a person familiar with identities of plants, the land is like a book; each change in the makeup of plant groupings is a statement about the history of that land, its soils and weather, and the subtle faces of the seasons. Inasmuch as plants are the most sensitive expressions of the seasonal clock, they can also serve as alarms. Sudden loss of health, an unusual inability to thrive, may well mean that an area is becoming unhealthy for all its residents, including people.

We are, after all, very much connected to the earth; yet our present culture does little to acknowledge this dependency. Many of us have grown up feeling that a knowledge of plants is primarily a concern for farmers and scientists. We eat plants, however, and we live among plants and plant products. Some plants are special to us. We see in them attributes and reflections of ourselves. Then there is that fascinating experience of finding a new plant, never before observed, and then realizing that it was growing all around us. Where had it been before? What else is there waiting to be seen?

Learning about the environment is learning to be aware. Complex ecological concepts are often based on subtle sensory distinctions. We begin as children to explore the world through our senses, building up internal images, codifying experiences with words and compounding the words into abstract notions. These abstractions become reference points to guide behavior, for an individual and for a culture. In other words, a person who has climbed a mountain to see the alpine flowers in bloom, a person who has taken pleasure in the summer flowers in the vacant lot next door, will want to protect those places, those experiences. A person who has never felt connected with a living land cannot be expected to care.

The activities that follow are based on sensory experiences. They assume the "botanist" is a beginner, either a young person or a beginning learner. Each activity has the potential for communicating a concept on the way plants live, but the main intention is to bring the learner in contact with the actual plants, alive and growing in the natural world.

2

Creating an Environment for Learning

The biology of plants is always the study of relationships: the relation of flower parts to each other, the plant's relation to pollinators or predators, or the interactions of plants as competitors or friends in a community of plants. The scientific terminology will seem difficult only if it is unfamiliar. For instance, we use the terms *chrysanthemum, rhododendron,* and *iris* in daily language, and they are botanical terms.

Any terminology is intimidating if it is not based on direct experiences. If a number of senses have been involved, the information base is larger and the concepts have more meaning. Therefore, the activities that follow are all experience-based, hands-on events.

Most of the activities have come out of the course records kept on the children's classes that were taught under the oak trees in a botanical sanctuary and garden called Garden in the Woods, the headquarters for the New England Wild Flower Society in Framingham, Massachusetts. Over a twelve-year period, a group of people, most of them volunteers, put a great deal of thought, time, and caring into creating a nature study program based on plant studies. Activities that involved native plants were favored. The group was committed to weeding out the activities that were just "busywork" and to developing activities that cultivated an appreciation of the plants. The activities were kept and reused if the children showed that they enjoyed them: either by their expressions of involvement, excitement, or pleasure; or by their ability to incorporate the experience in conversation or in an art project.

No activity will succeed on its own. Much of the success depends on the teacher's attitude. As we developed our skills at interpreting the activities for the children, we kept records of techniques that worked. In general, those records show that the more attention and affection we showed to children as individuals and the more we evidenced our pleasure in being outdoors with them, then the more successful were our lesson plans. An overreliance on materials, schedules, and right answers would tend to blind us to the children's natural abilities to enjoy their own responses and learn from them.

The following statements will give you some idea of our philosophy in teaching environmental awareness to groups. Even if you are teaching only yourself, or are dealing with older beginners, many of the observations will still apply to you. They all have to do with encouraging beginners to feel safe enough to open up and teach themselves.

■ Put yourself in the learner's shoes. Take into account what a child is experiencing, both physically and intellectually.

■ No one can learn when he or she is afraid.

■ Encourage exploration. Let your participants' guesses guide your questions.

■ Become comfortable with your subject. The more at ease you are with your basic objectives, the more you can let a class pursue a variety of paths toward that objective.

■ Teach by example always: go quietly, stopping to look and listen, smell things, handle living things gently.

■ Be as accurate as possible with common names. Use your knowledge to help others figure things out. If a group seems involved in the question and is on the right track, it is all right to leave them excited about the pursuit without providing the definitive answer. Acknowledge their answers as good possibilities.

■ Let children know that your information came from someone else and can be passed on by them with the same authority.

■ If the leader participates in the activity, it is important that the activity call for individual responses, not a right answer or best drawing.

■ Children are impressed by their own discoveries.

■ Stick up for your feelings and values, but don't expect others to respect them unless you can respect theirs and show it.

■ Many experiences not necessarily understood by small children are still valuable as building blocks for future knowledge.

■ Watch a youngster's body language. Learn the look of excitement, relaxed enjoyment, the flush of discovery. Share children's pleasure without overwhelming them. Learn as well the look of uncertainty, distrust, and confusion. Often these feelings can be diluted by a little individual attention.

■ Find out what kids already know about a subject.

■ Level with the group as to your intentions, your plans, and the schedule. Whenever appropriate, involve children in decisions and honor their choices.

■ A distracting youngster is frequently asking for some special attention. Ask the child's opinion or help in setting up an activity, or touch the child as you speak to the others. Try not to let the rambunctious activities interrupt you, as the whole group will then be distracted. Energetic kids are often trying to master an unfamiliar environment with physical activity. Provide active events to help them feel competent.

■ Watch out for heavy moralizing on the impending destruction of the natural environment. Fear of pollution and species loss can be taught easily, but if the concepts are overloaded with emotion, the end result is likely to be an aversion to learning more. Concentrate on the joys of experiencing what is there at hand.

3

When the Learners Teach the Teachers

THE JEWELWEED REACTION

The children's group had been standing quietly, listening to the leader's discourse. Their attention was directed toward the tender green leaves, orange flowers, and pressure-sensitive seedpods of the jewelweed, or "touch-me-not." The technique of touching the pod and exploding the seeds was demonstrated, and the children registered delighted surprise. Everyone wanted to pop a pod. The group's demeanor changed. Faces and voices showed excitement as senses became enlivened by the search. The children readily shared their enthusiasm and discoveries with each other. Some wanted to try to take a pod home to share.

It seems that of all the information that might be presented during the course of a guided walk, the fact that jewelweed seeds explode from their pods at a light touch became the clearest and best-remembered event. Giving the children a chance to teach themselves, and be successful at it, was a stronger statement than any assemblage of facts.

FOCUSING

Pocket lenses were passed out prior to an outdoor excursion. The children immediately bent to the wonder of the magnifying glass. All were involved, and one little boy seemed particularly enthralled. When the lenses were returned, his came back minus the glass. He hadn't lost it, he said. It came that way, but it still seemed to work all right. Sounded unlikely, but when the teacher tried it, she found that the remaining plastic circle did act as a lense in that her visual field was focused and intensified by the restriction of the frame. The resulting intensification seemed very like magnification! By experimentation, it was found that other senses could be accentuated in a similar manner. By "aiming" the sense first, looking for a particular color, listening for a kind of sound, the sense seemed strengthened and the impression was more vivid.

I.D. ETIQUETTE

Simple good manners can sometimes arouse a feeling of familiarity. One of our teachers recalled a childhood experience in which an instructor of a fern class required that students "say good-bye" to each fern, addressing each by name. Something about speaking directly to a plant, using the polite tone of respect, cemented the identities of those ferns into her memory.

We try to create a similar effect by requesting that students shake "hands" with various sturdy plant specimens, especially the evergreen trees. As we direct them to an awareness of textures in their hands, we remind the children of the usual courtesy of gentleness one extends to new friends. And then, there's that special feeling that comes from hugging a large tree.

WHAT YOU DO SPEAKS CLEARER THAN ANYTHING YOU CAN SAY

One teacher found she can elicit touching/smelling/listening responses from children just by making a point of doing these behaviors herself. In the middle of talking, she would suddenly make a point of observing a leaf bud or the texture impression of bark on her hand, revealing her own curiosity but not saying anything about it to the children. She found that they were soon imitating her actions, even unconsciously, and that this led to discoveries as they walked along together. Over a period of several weeks, this behavior leads to a decentralization of roles as the children take on much of the responsibility for finding interesting things.

4

Botany for All:
Special-Needs Children Out-of-Doors

Many of us who have special needs or require special care often have limited access to nature. If a trip through the woods or along the edge of a field is impossible, then a changeful garden or flowering houseplants of different sorts can bring pleasure and interest into a life bounded by limitations. The increasing use of plants and even animals in hospitals, nursing homes, and prisons attests to the healing and helping aspects of contact with natural processes. The following are only three images from our many experiences that resulted from our special effort to reach people with special needs.

On one occasion, a collection of plants and plant parts were brought into a nursing home and passed around among a group of people confined to wheelchairs. The goal was not necessarily to instruct the people, although any questions were answered, but rather we wanted to give the people a chance to feel the different textures of oak leaves, grasses, twigs with buds, and various barks. At one point, a piece of birch bark was passed to a woman who had kept her head down and had watched sullenly as the other items passed. At the sight of the white bark, her hand shot out. She held the bark, stroking it gently, and began to tell the instructor eagerly of her childhood love of the birch trees near her home. She responded similarly to other items as they passed. The instructor was later told that this was the first coherent communication that the woman had made in several months.

The second image is that of a boy with a hearing impairment, sitting high on the branch of a pine tree he had just climbed. As several of his peers climbed carefully past him, he looked down, smiling broadly, and motioned to his teacher. He signed to her that he wanted her to tell the leader, who was standing next to her, that here he was, ten years old, and this was the first time he had ever climbed a tree and it really made him feel happy. The gladness in his face made the signs and words unnecessary.

Not every experience with a special-needs group runs smoothly. People unused to the feeling of natural space can respond with fear and seek distraction by making trouble. Sometimes the cure is simply time. When a small group of mentally impaired adolescents came for a walk through the woods, it seemed at first as if nothing could reach them. They clung to their teachers, interacting only with them. The tickle of the pine needles and the coolness of the soil under the trees made them uncomfortable. We encouraged the teachers to touch things, and that enabled the children to begin their own search for textures. The greatest change came where the children had to cross the shallow brook on a narrow bridge. While each child had to be coaxed to make that first crossing, their excitement at success was so great that we all crossed back and forth several times. As we walked away from the bridge, one of the kids, instead of clinging to his teacher, gently pushed the teacher away and took the lead as the head of the group, moving eagerly from one texture of leaf or bark to the next, stopping often to share his discovery with the rest of us.

It should be obvious in all these examples that a confrontation with nature alone was not the special ingredient; rather, there was support for each person. It was clear that by respecting a person, a way was opened for further interaction with the world and its ways.

5

Structuring the Excursion

In the process of developing the learning activities, the teachers of the children's classes found that certain patterns and approaches were more conducive to satisfying the needs of both students and teacher. The following suggestions may help you create a comfortable structure for learning:

The pattern of short meetings at the beginning, explorations outdoors sometimes including a major art activity, and a discussion time at the end seemed to work well for most groups.

Try to put specific structured activities at the beginning of the class period and more open-ended activities that require individual approaches toward the end. This shifts the focus from teacher-centered activities at first to child-centered ones at the end.

A gathering at the beginning could be used to introduce a theme (with items to look at and talk about). You can find out what the children know about the theme and give them a chance to be part of the decision making or let them in on your plans. If you are working with a larger group of older kids, this is a time to get commitments from them as to what activities they will explore (subject to change, of course). This is a time for all of you to get a feeling for what is ahead so that all can proceed confidently.

Make sure the kids are not going to be distracted by uncomfortable clothing. Check that they have on appropriate shoes and clothes before venturing outside. Be ready to modify the trip if a child is poorly prepared, but don't make a big deal out of it. It is seldom the child's fault.

Set up clear standards of what can be collected. Limit the collection of live material to only essential examples. Try to create or modify activities so that nonliving material can be used.

During the exploratory phase, try to vary the pace. Include physically active games with set limits. ("Run to that oak tree and find a large leaf and a small leaf and run back to me.") When you get together for a quiet time, give the children a focus. ("What can you hear in ten seconds?" or "Imagine being very small and walking on this mossy stump.") If you have a short book on an appropriate subject, reading outside can be a pleasantly familiar activity.

When teaching new exploratory behaviors (such as taking apart a bud or a seed) proceed step by step as a group. This will provide time to absorb new impressions and learn new vocabulary. Such deliberate dissection also encourages group discussion as questions arise.

Encourage the children to work together to achieve goals, such as turning over large rocks or climbing a tree. For small children especially, learning to trust their peers is a new and important discovery.

If the same location is used for a series of visits, focus on different areas or use a different theme for each visit. It is also important, however, for children to realize that they don't know everything because they have experienced one visit. Focus on changes the second time around.

If you are going to use material gathered in the field for later activities, it is important to begin the activity soon after the collecting trip. A prolonged wait diffuses the interest in the materials.

The wind-up time is a chance for you to get feedback on what impressed the children. Ask them to remember just one thing they saw or did, or perhaps a favorite event. Encourage them to teach their parents or friends the things they have learned.

6

Up Your Sleeve: Devices to Enhance Learning Situations

Collecting bag—for items not attached or alive. Can even be used to recall sounds: Write impressions of the sound on a piece of paper that is kept in the bag.

Texture clues—an assortment of familiar items, such as sandpaper, cloth, toothbrush, and so forth, that can be compared to textures found on the way.

Color samples—for matching with natural hues.

Thermometer—to check for variations in temperature of soils or water: (1) up, down, and around a slope, (2) through a season, or (3) at the beginning and end of a walk. Some people simply enjoy working with measuring instruments.

Magnifying glasses—to examine the world of close-up. A nice group activity is to make a group circle, tummy-side down, around an interesting spot or object and discuss what can be seen without, then with, magnifying lenses.

Frames—rectangles of cardboard with a smaller rectangle cut out of the center, serving as a focusing frame for "picture-taking" and comparing the light and textures of different habitats.

Spread sheet—lay out a light-color sheet and check later for fallen debris. Especially effective during autumn.

A feely bag—a soft cloth bag or large sock in which to hide a mystery item to be identified by touch.

When exploring any new item, get the children in the habit of taking turns describing just one attribute they notice. The next round of turns might be the naming of something of which the object reminds them.

Also in your bag of tricks might be an example of a typical leaf or fruit that could be shown then searched for in appropriate sites. Checklists of items to find could be handed out to help structure a walk.

7

Individual Learning Styles: Different People Learn in Different Ways

Evidence from brain research indicates that creativity and artistic expression are possible in every person and can be enhanced by certain kinds of activities. One-half of the brain specializes in intuitive, spatially sensitive, artistic behaviors. These attributes are vulnerable to being overriden by the logical, verbal talents of the other half of the brain, especially as our current educational standards tend to reinforce the logical expressions.

Recent studies have suggested that there are fine distinctions within the constructs of logic/spatial styles of learning. These studies remind us that a child who can sweep us away with a recall of names and numbers may need encouragement in developing an artistic ability. Or the verbally shy student with a terrific sense of pictorial design should be rewarded for artistic excellence as well as encouraged in accurate verbal expression.

The activities in this book attempt to offer examples appropriate to every learning style, but it is up to the adult to set the balance between pride in natural ability and success in difficult areas.

8

The Art of Learning

Art projects are successful ways to involve a person in an idea. Children will gain from such projects to the extent that they have a choice in selecting the activity. Collecting materials and then using them in an art process that allows freedom to experiment will produce results that please everyone. A knowledge of the art process is an important element. The children can watch you perform the basic method (dyeing, printing, and so forth) and learn readily from your motions and descriptions. They will ask for help if they need it. Each child will approach a problem differently.

Environmental education has long espoused a learner-centered philosophy of education. Much of what a person experiences in nature supports notions of self-worth. The pleasing randomness of unique events and forms that the individual experiences is gradually ordered by learning about seasons, the characters of a species, and patterns of relationships. In an art project, a person makes intuitive and personal selections; the elements are experienced as related and important. The expressions can be understood to be reflections of an individual's interests and understanding. It is also clear that a finer appreciation of natural form and relationships produces the most interesting artwork.

Plants lend themselves nicely to this process with their differing forms, fairly passive states, and dramatic transformations. As children become familiar with new plants, these plants are more accessible as personal images. Art activities also demand sensory involvement, essential in developing descriptive vocabulary for distinguishing species. The ability to identify and separate species is the first step to an appreciation of the concepts of habitat, ecosystems, and life cycles.

9

Ideas for Art Projects Using Plant Material

Here are some of the ideas found under diverse plant topics described later.

DRAWING, PAINTING

Develop drawing skills as a means of making large items smaller and small ones large; create new, imaginary plants; make murals of "gardens" of everyone's plants; silhouettes of children get transformed into "plants," including adaptations to particular habitats; record growth of young plants with a series of exact-size drawings; use inks made from berries or smear tender leaves and flowers on paper for natural paints.

COLLAGES

Use natural materials or magazine pictures to make murals or individual plants or gardens; use waxed paper and an iron, or use transparent, self-adhesive plastic, to preserve and enhance a collage.

MOSAICS

Glue textural and colored plant elements on cardboard, sandpaper, boards, and so forth, to make pleasing designs. Also try seeds, beans, nuts with stones, sand, and soil.

WEAVINGS, MOBILES

Use criss-crossed lengths of yarn (naturally dyed?) to hold up a layer of found natural materials.

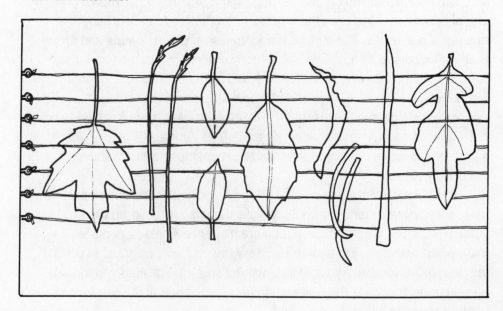

RUBBINGS

Capture texture rubbings by rubbing a crayon or soft pencil over a thin but tough paper, such as onion-skin paper. Metal foil can also pick up delicate textures for variation. The shape, color, and vein pattern of a leaf can be transferred to cloth by putting the leaf on a board, holding a piece of fabric over it, and pounding on the fabric where it covers the leaf. The print will stain through.

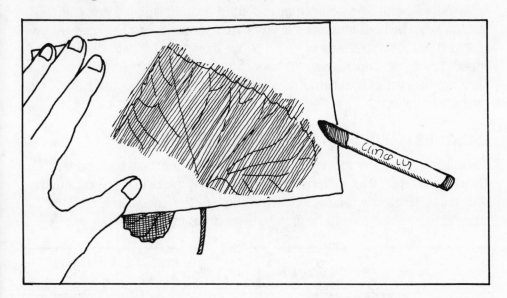

MODEL MAKING

Models of real or imaginary plants can be made from paper, flour dough, clay, or papier-mâché. An easy material for modeling can be made using a crust-less slice of white bread made of highly refined dough (about a tablespoon), enough white liquid glue to turn the bread back into dough, and several drops of food coloring. After it has been modeled and when dried hard, it can also be painted.

HABITAT HATS

Make some crazy hats, using paper plates for the bases and plant materials for decorations. For a more biological theme, call the hats "camouflage accessories" and use only material from specific habitats.

The habitat concept can be expanded by using long grasses and twigs as whiskers or feelers. If you are planning a night walk or are studying animal adaptations, have the children collect stiff and slender stalks to act as sensors to alert the wearer of nearby objects. If the stalks are attached firmly to a headband or armband, they can be impressively effective. Encourage experimentation and show examples of animals that have distinctive whiskers. Cats have whiskers on their elbows! This activity can be a lot of fun: people tend to look very peculiar with grass stems sticking out from their heads and arms.

PRINTS

Inexpensive printing blocks can be made from firm fruit such as apples, pears, citrus fruits, and especially potatoes, which can be cut into new shapes; leaves glued to cardboard can be inked and pressed onto paper; stencils can be cut using leaves for outlines, turning or flipping the stencils to make patterns; rubbings or tracings with a special textile crayon can be used to print T-shirts.

PRESSED LEAVES AND FLOWERS

Pressed for several days between newsprint to dry and flatten them, plant materials can be used innumerable ways. Arrangements can be preserved by sealing them between sheets of waxed paper, using a hot iron to close the edges. Try waxed-paper sandwich bags or layers of self-adhering plastic. They can be used as bookmarks, book covers, mobiles, name tags, card games, and so forth.

CREATURE MAKING

Use a firm vegetable, such as a carrot or potato, for the main body parts. Then attach other plant materials with glue or toothpicks to represent other body parts. Pinecones can be attached with pipe cleaners. Make a creature adapted to a certain habitat by using only material from that area.

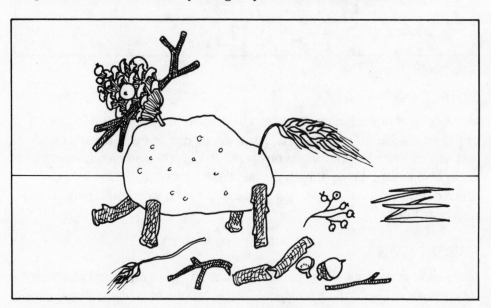

MASK AND PUPPET MAKING

Using plant materials, make a face mask or a puppet to create a talking plant or the spirit of a place. For a base, use papier-mâché, paper bags, socks, or old Halloween masks.

Many thoughtful teachers have observed children learning through the art process and have carefully documented their observations. The role of the teacher in enhancing (or inhibiting) the joy of learning is crucial. A clear and thorough description of this role can be read in Peggy Davison Jenkins's book, *Art for the Fun of It*. Most of the book, a vast collection of ideas and recipes, describes art projects for young children.

10

An Example of an Art Activity: Dyeing with Plant Materials

Pigments can be extracted easily from leaves, flowers, barks, and roots. The resulting dyes can be used to color wool yarns, which can in turn figure in simple weavings or more elaborate crafts. Simplified, the dyeing process could involve even younger children. In its basic form, it also could become an accessory to various other craft activities.

MATERIALS

Materials needed include a source of heat that can be kept at a low boil, saucepans made of stainless steel or enamel (aluminum will modify the color but can be used), a strainer for extracting the loose plant material from the dye bath, and a handled vessel to catch the strained dye bath and return it to the pan. A long-handled spoon or a stick will help in stirring and extracting the yarn. You will want to experiment with several kinds of plant materials, combining them for interesting effects. You will need to buy potassium aluminum sulfate, a mordant, from a craft store. You need a mordant to make the dyes stay in the wool yarn. Different mordants make different colors from the same dye stuff. Drugstore alum will do if you can't find any other. Get white yarn, the cheap kind if you're just playing around, but it must be pure wool. Cotton or acrylic won't accept the colors.

PREPARATION OF THE DYE BATH

Fill a quart or half-quart pan two-thirds with water and put in a handful or so of plant material. If you are using nut hulls or bark, soaking the material overnight and then heating it will speed up the process. Simmer the material, just under a boil, until the water is tinted by the pigment. If you are using delicate dyestuff, such as leaves or blossoms, it will take fifteen minutes to half an hour. Woody material may take an hour. When the water seems as colored as it can get or you can't wait any longer, strain out the plant material. The quickest way is to pour the "soup" through the strainer into another container, then return the dye bath to the pan. Put about 1/2 teaspoon of mordant into the bath. The color might change somewhat.

The next step is to introduce the wool into the dye bath. When cool wool is thrown into hot water, the fibers shrink and mat. If you are using expensive wool, you need to let the bath cool and to presoak the wool in warm water. With cheap wool, we just throw it in. First make sure that it won't get tangled with itself and other samples by wrapping the wool around your fingers to the amount you want, to make a loose "hank." Then use a long end to make a simple knot around the hank. Make the knot easy to undo, as you may want to use the wool while it is still wet. Short pieces of string, tied loosely around the hank, will also keep it from tangling. If you are working with a group, that long tag end of yarn is important. The tail can be draped over the edge of the pot for easy extraction or labeled with its owner's name.

Start checking on the color of the yarn almost immediately after putting it in. Pinch the wool to see the hue of the dye. The longer the wool sits in the dye bath, the darker it will get (up to a point, of course). Use a series of hanks to get a range of hues. As you wait for the dye to set, the heat should be on a low simmer, or the pot can even be off the heat if you are concerned about small children working around a heating element. A hank can be extracted from the bath and rinsed in fresh water to stop the dye process at any time. You will notice that the colors look much lighter once the wool is rinsed. The hank can be put back in the dye pot, or you can skip the rinse step if the wool is to be used for simple weavings such as God's-eyes.

You may want to try altering the color of the dye with other chemical mordants. The presence of iron (old iron nails or iron pills) will darken the hue, called saddening in dyeing parlance. Cream of tartar, a pinch or so, can brighten a color. A whole range of colors can be found in the vegetable pigment anthocyanin. If you are using plant materials that give off a red, blue, or purple color, you can play around with the colors by adding either ammonia or vinegar to the dye bath. (Anthocyanin is sensitive to changes in acidity.) Dip parts of a hank in a solution for multicolored yarn. None of these colors will be truly fast; they will eventually fade in sunlight. For small children and simple weavings, we recommend not even rinsing the wool. Just wring out the dyed hank, then blot it.

The children's favorite dyed yarn projects allow for immediate use of the wet wool. The wheel weavings known as God's-eyes are the most popular.

Even the process of neatly wrapping a stick or a stiff card with various colors of yarn is satisfying to many children. The colors really are lovely, especially

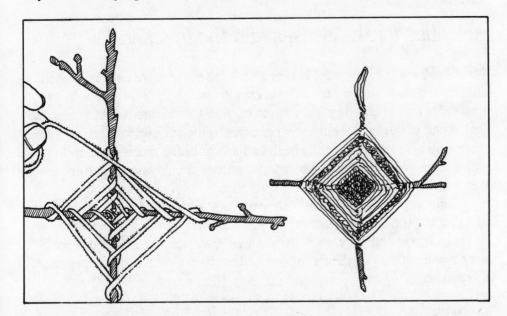

when fresh from the dye pot. A dyer invariably gains a greater appreciation for both the bright colors we commonly wear and the subtle varieties of hue that can be distinguished as people learn to work with colors.

It is recommended that you do most of the dyeing in the autumn, when the plants are at the end of their growth cycles and collecting is not so damaging. There is also more material available at that time. All seasons have their colors, however. The same dye plant will produce different colors at different times of the year. The following are some suggestions for strongly colored and easily available materials.

Yellows (by far the most abundant)—Any of the daisy family: marigolds, goldenrods, sunflowers, black-eyed Susan, dandelions; dock roots, birch leaves, onion skins (one of the best sources), and many more.

Browns (most nuts and barks)—Black walnuts, coffee grounds, tea leaves, acorns, pecan shells, onion skins (using a longer time in the bath).

Greens (very rare, as the green chlorophyll in leaves is destroyed by heat)—Horsetails or birch leaves in spring give a yellow green, poverty or broom grass (*Andropogon scoparius*) gives a green in midsummer.

Reds (a permanent red is rare)—An insect that feeds on cacti, called coccineal bug, gives a good red as do the roots of the bedstraw family, principally madder.

Red-blue-purple (colors derived from the plant pigment anthocyanin)—bases (ammonia or baking soda), but none are permanent colors. Purple cabbage leaves, blueberries, blackberries, strawberries, beets, pokeberries.

Try some of these common plants first, but experiment with materials in your own area as well. You are sure to find new colors.

11

Activities for the Senses and the Imagination

We all have felt the difference between knowing because we have read about a subject and knowing through a direct experience. There is a fullness in the knowledge that is based on sensory experience; the information becomes integral, a part of our lives. Sensory experiences in the out-of-doors can revitalize those senses of hearing, sight, smell, and touch. In the process, the individual gains a new confidence. Any educational activity should offer opportunities to increase both self-knowledge and information.

Some of these activities are "shorts." They are ways of beginning a class and drawing the children together, or you may find them useful as breaks or as peaks for closing the session. In all these examples, the material is not nearly so important as the feelings evoked or the self-affirmation experienced by each person.

FRAMES

Magnifying glasses are the ultimate frames: they concentrate the vision and focus the mind on some small part and make it seem important. We see it anew. Close up, it is more interesting than we thought. The magnification is not the most important part. Using the frame of a filmless slide, a sheet of cardboard with a hole cut in its center, or even peering through the aperture of a loosely held fist will eliminate the peripheral distractions sufficiently to give the impression of close-up magnification. Frames are good tools for investigating a variety of textures. They are also useful for comparing the color, light, and textures of habitats, especially when the viewer can stand in one spot and frame different views in succession.

COMPARISONS

Any sensory information is enhanced by comparisons. Experience different textures in rapid succession (using one hand, feel rough bark, then smooth bark soon after) or touch something rough and something smooth at the same time. Sense the brain's attention move rapidly from one sensation to the other, never both at once. It is also interesting to notice how quickly our sense of touch is lost when the hand rests on a texture for a few seconds. The smallest movement renews the textural information. Without comparisons and contrasts, we would be numb.

NEW WORLDS

Small Worlds

Find small worlds within larger worlds. Think of a patch of moss as a forest of giant trees. What would it be like to be a tiny creature climbing up a stump as if it were a mountain? What is it like for a bee to enter a colorful flower? Put your eye right up to the flower to see. See a forest as a giant's patch of moss.

Take a Question Walk

Make a list of questions only. After collecting a small sample of questions, have the children suggest possible answers. Help them out with answers if you want, but present your information as being based on your guesses or on answers you have heard from other people. Let the children decide if the answers are "good" ones or not. Do all you can to emphasize the children's abilities to ask questions, even ones that can't be answered. If later experiences provide answers, remind the group of the question and identify the questioner.

Make the Familiar Unfamiliar

Look at a familiar view upside-down through your legs. It really does look different, especially if the view has depth (down the path, across the field, out the door). Explore an area as closely as it can be seen. Each person will have a different focal length. Walk backward down a wide, even path. Look up a tall tree slowly: start with the ground level, think about the water moving into the roots, then "watch" the water move up the trunk, along a branch, out a twig, a stem, and into a single leaf. The next leap would be out the underside of the leaf, as water vapor, and rising up into the clouds.

Be a Bean

After the children have had some experiences with growing plants from seeds, let them imagine being the seeds they have studied by acting out the various stages of plant growth. Include events you know they have seen. These are some possible stages: sprouting, growing leaves and branches, getting taller, feeling a gentle breeze, then a strong windstorm, getting ready for winter, then the cold, awakening in the roots and buds in spring, getting old, falling to earth, and, finally, decaying and becoming soil.

Two Worlds

This activity can expand into a very elaborate program if you wish. The children will be creating the natural history of two planets with different climatic conditions. One planet can be hot and dry part of the year, then flooded with water the remainder. The second planet can be arctic in character, with a warm growing season and a very cold, dark winter. Have the children form two groups, and let each group work out the various plants and animals, food chains, and the humanoid cultures that have adapted to the environmental conditions. The groups should work independently, even secretly, to add drama to the second phase of the activity.

After the planet's ecosystems are worked out to the groups' satisfactions, let it be known that the arctic planet is growing too cold to support life and the humanoids must send out space probes to discover a suitable planet to colonize. The two cultures exchange material, decipher each other's messages, and try to decide if the life forms are compatible. The arctic people must decide from the samples they get whether they can survive on the other planet. The people of the stable planet must decide whether they would be threatened themselves by the new life forms. The issues and concepts involved provide excellent discussions for middle school aged children and older.

12

Inventing Seeds that Can Travel

Young plants benefit from growing away from their parent plants: sun space and soil nutrients will already be depleted where the parent grows, and a new area might have better resources. The following activity asks participants to design "packaging" that both protects the seed and makes it possible for the seed to move to a new environment. Include in the design a means to move to an environment where the plant can thrive. For example: seeds from plants that grow on the edge of a pond may be small enough to stick in the mud on a duck's foot and get carried to another pond edge where they can grow successfully.

With young children, make sure they have some experience with the concept of traveling seeds. Provide them with seeds with fluff (dandelion, milkweed, cottonwood trees) or helicopters (various maples, ashes) or hitch-hiking stick-tights (burdock, beggar's-ticks, tick-trefoil) or some fruit with hard pits that birds eat (cherry). Your area may have plants with water-floating seeds, as well. Check stores that carry material for dried plant arranging for unusual examples. Having played with and examined the structures of such seeds, children will be better able to design their own seeds. You might introduce an element of competition (as there is in nature) for the creation of a seed "ship" that takes a seed the greatest distance or floats it for the longest time. Encourage the free exchange of ideas in the initial planning. Point out that experimentation with even unlikely possibilities is another principle of nature.

The "seeds" to be moved could be any objects of the same size. Beans are easily obtained at the supermarket and continue the theme of seeds as precious packets of potential life. You might introduce a more dramatic elaboration for children who are familiar with animal migrations and other seed journeys. Create an involved fantasy about the recent appearance of an island, not far off a main continent. Let the kids organize themselves into groups that are responsible for inventing new ways for plants, through their seeds, to populate the new habitats. They must consider the usual modes of transportation and then create a given number of seeds to represent each style of dispersal. Set aside the time for each group to report on their results and demonstrate and describe models. Consider having the seeds "themselves" report on their individual journeys, either orally, or with illustrated narratives.

For discussion

■ Look at pictures of different habitats of the world and make guesses as to the predominant mode of seed transport in the area. Choose extremes for contrast: Sahara desert/Pacific islands, arctic tundra/Amazon jungle, and so forth.

■ Does sending its offspring away offer any advantage to the plant that produces the seeds?

13

Ways to Exercise the Senses

Many of the following activities can be used as part of any of the plant topic activities that come later in this book. Some are further developed in that section. Some children enjoy using these activities over and over, making up their own names to identify the activities.

■ Pretend to be a specific animal, searching for its dinner: a nuthatch or woodpecker on a tree, a squirrel or rabbit on the ground, or a heron or hawk patiently watching for small movements.

■ Follow with your eyes the imagined journey of a beetle up to the tip of a branch. Try to see the path through its eyes.

■ Estimate distances, weights, heights, and temperatures, then use the appropriate instruments to find out how close you came.

■ Measure a number of things using your finger, arm, foot, or height as the unit of measure. Or list all the things that are the length of your finger.

■ Take apart a block of soil, a bud, or an acorn. Dig into the earth. What are "insides" like?

■ Moisten under your nostrils and try smelling what you smelled before. Hold something cupped in your hand and breathe moisture on it before you smell it.

■ Using your hand, find the coldest place and the warmest place in an area. What factors might have made the difference?

■ Run an evergreen needle or a leaf edge over your upper lip or the side of your face. Which skin is more sensitive? Which parts of your hands give the best touch information? Swing your hands through the air, to the front, then to the back. Which surface is most sensitive to temperature, the back or the palm of your hand?

■ Develop a common vocabulary for textures, then play guessing games with various found objects. Children can find objects to put into a bag for others to guess, or they can describe the objects by texture and let others guess.

■ Designate special spots for sitting quietly and listening. Close your eyes or use blindfolds if possible. This is sometimes difficult for very young children. Holding hands in a circle may help relax the group. Are some sounds more pleasant than others? Do different trees make different sounds? Put your ear to the trunk of a tree when the wind blows.

■ Give each person a leaf and suggest the children try to think up as many ways as possible to make sounds with the leaf. Do leaves of other kinds of trees make different sounds? Compare with the fallen leaves from last summer.

■ Drop familiar objects where they can't be seen but can be heard. What are they or what do they sound like? Look for other objects that will make similar sounds.

■ Draw a sound. Does it have a color, as well? Make a list of sounds in one area and compare them with someone else's collection. Compare with sounds from a different area.

■ Make a mystery tape of sounds from an area known to the students and play it for them to guess. Let them make a tape as they walk around. This is a good way to get otherwise active children to start listening.

■ Make a group of sound boxes. Each child finds an object to fit into a certain size of box. Tape the box shut and take turns guessing what is inside.

■ Draw a map of a walk, showing special discoveries and events. Compare maps among the members of the group. What events were the same, and which ones were different?

■ Find a home for yourself. Be yourself or a particular creature and find a place that has food, water, and shelter from predators and weather. (Find and mark an area as your own. A flag for marking can be made before going out.) Give the rest of the group a tour of your place.

■ Look for examples of just one color on the walk. Look for a spectrum of colors.

■ Make a list of the nouns and their verbs that describe things you see as you walk around. Later on, mix up the nouns and verbs for some amazing images.

14

Enhancing Observational Skills

LISTENING WITH "DOG EARS"

When we look at an attentive dog or cat, we see the ears pricked forward, catching and cupping the sounds back toward the inner ear. We can experience something of the same effect with cupped hands, placed behind the ears, bending the ears gently forward. Small children may need a little extra help getting the hang of this. Mentioning that the cupped hand will be "catching" the sounds the way sand is scooped up and held will help them keep their fingers together. Encourage kids to experiment by turning their heads back and forth or by turning away from the sound. Choose a place where the sound is varied and extended. The sounds of running water and wind in the trees are both very dramatic.

FINDING ABSTRACT SHAPES

Make a set of cards (index cards are good) of abstract shapes, one shape to each card. Possible shapes to try:

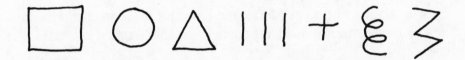

Give each child a card, or have the kids choose their own. If you are offering choices, have multiple copies of each shape. The shape is understood to be hidden in objects and can be found by looking and touching. Encourage children to share their discoveries as they go. This will help those who have difficulty with abstraction to see how it works for others. Participants may learn to team up when they see that some objects can be used for several shapes. Encourage their helping each other so that many shapes are discovered. Spend some time at the end of the walk recounting favorite findings.

NATURAL METAPHORS

This activity is a good stretcher for both imaginations and vocabularies. Put any one natural object up for inspection. Pass it around or put it in the center of a group to be handled. Encourage the kids to say what they think the object resembles. Whatever anyone says is right for that person. Respond positively to every analogy to keep the ideas coming. The focus can be shifted when ideas come more slowly. Point out the shape, change its position, ask about the color, texture, size, or sounds made by touching the object. Try adding a second object and explore its attributes for a child. A nonliving object is an interesting contrast with a living one. Then suggest that the two objects might have certain things in common. With two or three objects, the physical distance between them will alter the relationship, as will their relative heights. What happens to the relationship if one or both are moving in some way?

15

Centering

Centering is the process of setting down and being aware of one's feelings and thoughts. Taking a quiet walk or waiting patiently in a natural setting can lead to a pleasant sense of connectedness and discovery. The thing discovered may be a flower or a view, or a wild creature may come magically near, but the feeling discovered comes from within and is often accompanied by a sense of inspiration.

Poet Gary Synder has suggested that the first meditators were probably hunters, waiting alongside a game trail with relaxed alertness. An open, relaxed, and alert mind is very sensitive to its surroundings, very ready to perceive a subtle change or a new form. You have only to spend an afternoon outdoors hunting for pictures with a camera to feel that sense of expanded awareness.

To understand the value of centering, consider the characteristics of a person swamped by uncentered feelings: anxious, distracted, and prickled by "shoulds." The just-as-it-is quality of nature is an appropriate antidote. Becoming receptive to that quality means quieting the wordy ways of the explaining mind with the activities of sensory awareness, deep and relaxed breathing, and gentle body movements (if any).

Children can experience that awareness within the safety of a group. Left on their own, in unfamiliar territory, young children are likely to experience feelings of abandonment (as may any person left alone in a new place). With a young group, begin with short listening exercises, perhaps just a silent count of ten. Signal the beginning and end of the listening time by raising and lowering your hand, but as the children gain confidence and if there is something to hear you can extend the count. If everyone's experience is valued by the group, attention will turn increasingly to a general awareness beyond the group. Eventually the children will want to be farther apart in order to see and hear different things. Other goals of finding particular materials for individual projects or listing personal observations as poetry will strengthen the value of the time spent alone. If it feels appropriate, ask for the children's impressions of how they felt when "nothing was happening" and help them to see the value of their own thoughts and to appreciate the process as a time for getting in touch with the feelings and perceptions. That's centering.

16

Approaches to Centering

Have the children select a plant with which to spend some time. They can bring back to the group a sample of nonliving things they found, or a drawing or a list of what they observed as they sat with their plant—perhaps an imagined story of what a day or a year of being that plant would be like.

With young children, you might ask them each to find one special attribute of several different plants, or even habitats. Don't try to hold them down if they get restless. Once it gets to be a struggle, you know they are distracted by your demands. Just focus on their ability to notice new things.

With older children who are more interested in group activities, the group can form a "circle watch." Sitting with their backs to the center of their circle, each child looks out at a slightly different view. Each can then report on the different aspect that they see. Go around the circle several times to get a full picture of this "encompassing" outlook. If you write down the descriptions, the children will become conscious of their phrasing. Put each observation on a separate card. Work together arranging the phrases in a pleasing way to make a poem. Certain phrases can be repeated or an opening statement can set the tone (such as "We look out and see . . .). Read the phrases that describe a particular direction and let the group guess the direction.

One of the many gifts given by Native Americans is an image of a people living closely and reverently with nature. Many children can identify with this image of an "Indian" as representing knowledge, alertness, and respect for nature. Discuss how people who live closely with the land feel that each place is overseen by a particular spirit that must be treated respectfully. Disrespectful treatment (killing too many animals) would make the spirit angry and food would be withheld (reproduction among the animals would be low). Ask the children to use their feelings and imaginations to seek out some sense of what the spirit of their favorite place would be like. The result could be a drawing, a poem, or a mask made from gathered materials. What kind of behavior would the spirit approve of or dislike? Let the maker be the spirit's voice.

17

The Learner as Teacher and Vice Versa

Who learns more, the teacher or the child? Probably the one who does the teaching. There is something about being responsible for another person's awareness that considerably heightens one's sensibilities. On being asked a question, one often discovers a new answer in oneself.

Here are several ways of helping us as teachers to let children teach us: Put learners into teaching roles. As you near the end of a nature walk, ask the children to select some special leaf, give it a name of its own, and take it for a walk . . . to make up a leaf-sized nature trail. Then the child takes the leaf and the teacher over the trail and shows them items of interest in nature. Not only are you able to see the effects of your teaching, but the children can put their slant on the new information, making it their own.

Have the children choose special sites to interpret for others by writing a paragraph describing special plants or relationships found in the area. This can be done spontaneously as you go along. Stop at a likely spot and ask individuals what they would point out if they were a tour leader. Tie-on cardboard tags make good instant trail labels.

With young children, we ask them to remember "just one" thing they saw on the walk. An opportunity to express their experiences both emphasizes the value of group discoveries and primes them to share their experiences with others. Once they get going, there tends to be a flood of memories. Almost always, they remember information only when it is attached to something they discovered or did.

18

To Have and to Hold

Smooth seeds, special stones, a dried and flattened flower: these are mementos that keep us in touch with an event or a place. Those things provide a thread of ownership, a path by which we return to a time of special meaning. Collecting is a natural compulsion we share with squirrels and jays. For us, it is a process by which we select from our environment that which has meaning for us. It is our way of finding ourselves outside ourselves. In this process of discovery—whether it is finding just the right flower for an arrangement, recognizing an edible fruit, or searching for just the right spot to sit and dream—our senses are most open to our relationship with our surroundings.

Collecting without destroying is an art in itself. Here are three different ways to collect.

1. Instead of taking things away, put things in. Select the proper place to plant a young tree, or discover the right-feeling spot for a home while pretending to be a wild creature.

2. Developing empathy helps to establish personal connections. Choosing a certain tree to watch, draw, or write about gives the imagination a means for identifying with natural processes.

3. Collect by listing. Sounds, smells, and colors can be captured in this way.

Other activities that utilize collections can be found in this book by looking in the index under "collections."

19

Scavenger Hunts: Creative Collecting

Learning is greatest when impersonal facts are put in personal order. A scavenger hunt is a simple set-up for the exercise of expertise and creative problem solving. It allows children to affirm what they know, take some chances, and learn something new. Included are a few of the hunts we've used. Competition is a secondary challenge in these hunts. The expectation is that with each individual solution, the hunter will find out more about his- or herself.

■ A Scavenger Hunt for Sensory Awareness and Emphasis on Relationships

■ Find three brown things and arrange them in order from light to dark.

■ Find three different shapes (such as circle, triangle, and square) formed by the shape of the object and three shapes formed by the spaces between objects.

■ When the leader says "go," stand silently and listen. Be ready to describe or mimic the sounds you hear without using their names.

■ Find something that to you has a "happy" look to it and something you would describe as "fierce."

■ Find two plants growing on another plant and two plants growing on a non-living thing.

■ Find two animals on the same plant. How does the plant help them? Are they helping the plant?

■ Find a plant being shaded by a plant while shading another plant. Which plant do you think will be standing there in twenty years?

■ Find something that is changing back to soil. How many animals and plants can you find involved in causing this change? Are there other factors influencing the change?

■ Within twenty steps, can you find a place where you can demonstrate three temperature readings that vary by more than five degrees from each other? Are there different plants in the three areas?

■ Find an example of erosion. How do you think it was caused? Do you have any ideas of how it could be controlled?

■ Find something soft in the reach of one hand and something rough in the reach of the other.

■ Find some food that would be good for a squirrel; for a rabbit; for a deer.

This hunt would be a lot of work for nine-year-olds and up, so be selective in asking younger children to use abstract ideas. You might want to try out a challenge or two with every walk you take. Or use the challenges within the context of a treasure hunt: go to the big oak and do---, then go to the brook, and so forth.

A scavenger hunt can be given the added dimension of additive math problems. With every successful discovery, you can request that children add or subtract a number. If the hunt is sequential, specific numbers of paces can be part of the instructions. The winning team would have to answer all the challenges as well as have the correct final sum.

20

Developing Environmental Values

We all have feelings about places we visit. There are things we like about a view; spaces or colors or shapes that excite us or calm us down (or don't interest us at all). Describing our feelings often enhances the experience; the experience becomes "ours," distinct from the experiences of others. Owning an experience brings with it territorial feelings of wanting to preserve a place that we enjoyed.

For a group of young children, this process of owning an experience can begin with the question "What did you like best about the walk?" Older people can draw further from their memories: favorite vacation areas, quiet places around their homes, or fond memories of "a hiding spot" in their childhood. To speak of or to draw pictures of the remembered spot is to begin to put concrete form on elusive environmental values. After a group sharing of personal memories, introduce the question of what general values link all the individual experiences. What makes a situation special and memorable?

The vocabulary that originates from these discussions can be used to describe other places the group visits. Encourage the group to choose nearby favorite spots to go for special events. When an area becomes recognized as special, ask the children if there are ways the area could be enhanced or maintained. What would happen if lots of people decided that it was their favorite spot? These discussions, the vocabulary used, and the rationales that develop are important in building a sense of stewardship. The concepts and terminology are applicable to national parks and backyard hiding places alike. For older children, bring in some literature on environmental advocacy. Essays by John Muir, Aldo Leopold, and Rachel Carson are classics and are still pertinent. Suggest that children try writing about their favorite areas as if the areas were being threatened and needed defending.

Find out where local conservation areas are located. If a map is available, discuss the ideas about town-owned natural areas, how the land is acquired and maintained. Find out if the children think there should be more wild land, and how wild they think it should be. Should open fields be allowed to return to brush and woods? Should everyone have access to it for any purpose? Is the land accessible to walkers? Is it in areas where animals concentrate, such as wetlands? Is only one kind of plant habitat represented? Give the children a chance to plan their own town, to imagine that they were the town planners back when their area was first settled. What decisions might have been made to make the town land attractive to native plants and animals? What might be done now to provide for future situations?

21

Do You Have to Know the Names?
A Rationale for Correct Identification

The name that identifies and distinguishes a plant is a verbal condensation of the experiences of many people. A name is a unit of communication, our access to distinctions and uses. A name is the connection between the life of the plant and the lives of people. The common suffixes "wort" and "bane," for instance, are tags that label the plant as "good for something" (motherwort, liverwort) or "bad for something" (fleabane, wolfbane).

Many names, especially the Latinate scientific names, serve to organize like plants into groups. They may also describe distinctive attributes. The names are important, but they are memorable only if they are linked to personal experiences. The experience could be a special memory of a place in which a flower was first found or a plant's interesting use or a peculiar form or relationship with its environment.

Use the names. They literally are handles for carrying the plants around in memory. But make sure that the names are a part of the experience of meeting a plant and not masks blocking the way.

22

Activities for Identifying Species

GAMES WITH TREE IDENTIFICATION

The idea that different shaped leaves correspond to different species can begin simply. When outside in a wooded, leaf-strewn area, ask each person to find a special leaf and bring it back to the circle the group forms. The chosen leaves can be matched using various attributes: colors, textures, sizes, and outlines. Be open to attributes children might notice.

Introduce the common names of the trees whose leaves they have found, giving one attribute you notice that identifies a group of leaves as a species. Find out what the children observe in the way of likenesses.

Show the group one of the trees from which their leaves originated and tell them how you know. Some children need to see this selection process, as the green leaves may look significantly different from the leaves on the ground. Be sure to use the same vocabulary that the kids used in describing their leaves.

Encourage the children to find other species by their leaves and help them if they ask. Other parts of the experience might include discovering and investigating other trees of previously identified species. This can become a part of the rest of the expedition, but let it drop if other things become more interesting.

Another game can be organized in a more formal, structured manner by bringing a number of leaves from several species into the class before an excursion. Giving each person a leaf to be matched with other leaves is a means of sorting the group into smaller units.

Literate children can then list attributes of their species on a card, recording each suggestion. Drawings, of course, are also encouraged. The leaves are then returned, cards are exchanged among groups, and the players can then try to match the leaves with the descriptions on the cards.

When the groups are back together, ask the children which words helped them make distinctions and which ones were too vague. This is a good way to introduce some of the botanical terms found in field guides. Mention some appropriate new terms and show examples or illustrations in a flower guide.

Have children put their new skills to use by encouraging them to make a field guide to some of the plants in your area, grouping those plants that can be described as similar in order to distinguish them easily. Each child or group might take a single plant and describe its identifying characteristics. To keep a game feeling going, the descriptions can be exchanged, or put together as a scavenger hunt.

Another game involves using leaf gathering and identifying to get an idea of populations. Have children gather a specified number of handfuls of leaves at various points along a walk. Sort the leaves and count the number of species as well as the numbers of leaves in each species. At the end of the walk, go over the numbers and help the children see that certain places were, perhaps, "oakier" or had more willows or pines than others, based on the collections. Do differences relate to habitat conditions?

Here is another way to make the point that all leaves are different: Ask each child to choose a special leaf and perhaps give it a new name. After studying the shape, colors, and textures of their leaves, the leaves are placed in a pile and mixed together. Ask the children to try to find their original leaf. Surprisingly, this is easy to do even if all the leaves are of the same species. The exercise confirms for the children their ability to make subtle distinctions and impresses on them the uniqueness of entities in nature.

23

Developing a Vocabulary for Appreciating and Identifying Plants

Bring in a large assortment of plant parts: leaves of various shapes and textures, fruit, bark, roots, seeds, flowers. Ask for words that describe the textures of the things; encourage exploration of the objects in order to find many words. As the children suggest words, write them each on separate cards. When the suggestions are all recorded, go back over them as a group and see if any of the words mean practically the same thing. If so, write them on the same card. Let the children make the final decisions on each word. Mention that these words could be used to describe the objects encountered on the day's walk. To make it especially challenging, the objects that use two descriptive words will be brought back (if possible).

Set up a game board with the word cards running down one side, and a duplicate set running along the top. (See illustration.) Have the kids help. Tape or pin found objects to the appropriate station at the intersection of a side word and a top word. Try out a few items to acquaint the children with the process. Then hand out collecting bags and go out. (You may find it more

	PRICKLY	SMOOTH	SPONGY	ROUGH	FLUFFY	FLAKY	FURRY	BU
PRICKLY	X							
SMOOTH		X						
SPONGY			X					
ROUGH				X				

convenient to use this activity as a way to use items already collected.)

When you return and get involved in sorting and placing the objects, be ready to make new cards and to reorganize some of the groupings of like words into new categories. Children in their early years of language acquisition will readily absorb and use new words, even botanical terminology that seems difficult for adult minds. Have a few appropriate biological terms memorized and add them to the list. If you use the words in your discussion, the children will pick up on them easily. After the terms become familiar, read a passage from a botany text and let the kids be impressed with their knowledge.

A FOLLOW-UP GAME

You or the children can assemble a number of texture objects that they have previously observed and named. Keep the objects hidden. Put one object in a soft bag, secure the top, and let one person feel the contents and "broadcast" the texture adjectives to the other children, who try to guess the identity of the object.

24

Creating a Key for Species Identification

The process of sorting plants into piles can be used to introduce the concept of a key. A key is a mechanism for making distinctions among groups of similar plants until comparable attributes are sorted and related to the various species. The structure of a key is treelike. Each distinction is a branching that leads to further distinctions (branches) and that ends in the identification of a species.

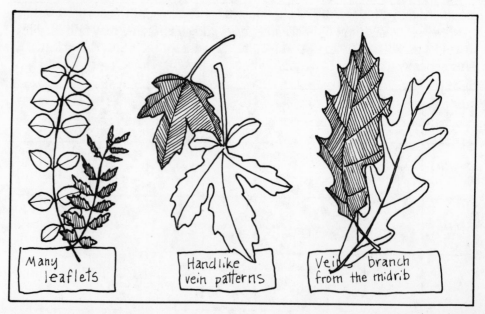

Many leaflets

Handlike vein patterns

Veins branch from the midrib

To teach the use of a key, make one as a group. Assemble at least twelve or fifteen leaves of local trees. The sorting process will be easier if several members of a plant family are represented. Using large sheets of paper on which to pile similar-looking leaves, first have the leaves sorted into general categories. All the leaves in a pile must share the same distinguishing characteristic but can be different in other ways. Evergreen and deciduous are easy first branchings. Use a third pile only if it leads to an identification within one or two branchings. Keep sorting the leaves into paired distinctions until all the branches end in an identification. Use new sheets for each new branch. Record the distinctions on the sheet. It will go faster if the leaves are already identified with labels, but you may want to challenge older children who are somewhat familiar with plants with a group of unlabeled leaves.

MAD HORSE

One of the first distinctions made in a key for deciduous trees is the differ ence in branching patterns. When buds and leaves are arranged in alternating steps along a twig, the tree has alternate branching. In opposite branching, the buds and leaves are opposite each other, like arms on a body. By knowing the relatively fewer species of opposite-branching trees, identification can be greatly simplified.

For the beginner, the term "mad horse" is a handy acronym for the families of trees with opposite leaves: maple, ash, dogwood, and horsechestnut. This series applies only to trees. A number of shrubs, including many introduced species, have opposite branching patterns. Be careful, also, to refer to only the newest growth of twigs and leaves. Older branches have been altered by environmental events and will have lost the original symmetry of their forms.

Activities for Learning About Plants

25

Seaweed Prints

Fresh seaweed has a natural gelatinous coating that acts as a light glue to hold the drying strands to a paper backing. The larger species may need a little extra paper glue, but the finer, flatter species will stick in whatever position they are arranged. By far the most effective way of arranging seaweeds is to float them in a shallow baking dish or pan and then slowly bring a stiff piece of paper up under them. Position the seaweed while the paper is just

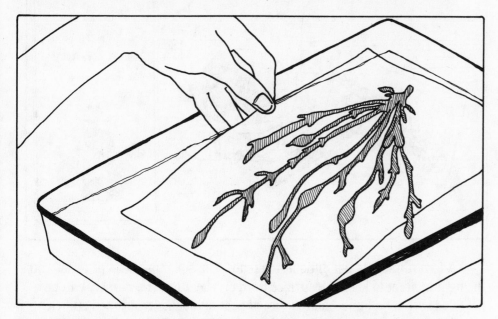

below the plant. Once the seaweed is lifted by the paper from the water, it should not be disturbed. Allow the sheet to dry a little, but before it is completely dry, press it between unprinted newsprint pages weighted by a book to flatten the sheet. A piece of waxed paper over the seaweed will keep it from sticking to the newsprint.

26

Fern Activities

Which are the fiddlehead ferns? All ferns go through a fiddle stage. As the fronds unfurl from a tightly wound coil, the central spine and compressed leaflets resemble the form on the head of a fiddle. The edible fern that most often appears in markets as "fiddlehead fern" is the Ostrich Fern. The rapid growth of ferns in spring makes all ferns, especially the larger ones, interesting to watch and impressive to measure for their daily growth. Children can measure the ferns against their own bodies. After a couple of weeks the larger ferns may outgrow the small children.

Ferns propagate themselves by underground stems and through dust-like spores produced by the fronds. Usually the spores appear as dark clusters on the underside of the frond, but some ferns have separate fronds or leaflets just for spore production. Spores are so tiny they seem like powder, but in the right environment, each spore has the capacity to create a new fern plant.

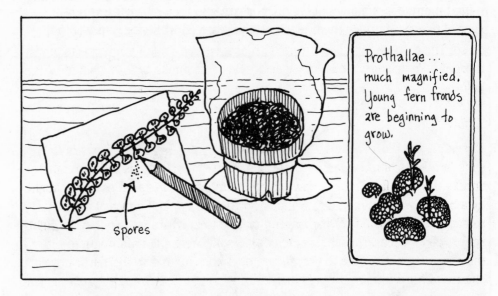

Prothallae... much magnified. Young fern fronds are beginning to grow.

spores

A fern nursery is not difficult to create. The soil must be kept moist and the light indirect to keep the surface from overheating. The nursery can be a pot filled with a half-and-half mixture of peat and sand, moistened sufficiently to feel moist to the touch. Tamp the soil with a block to level the surface. Sprinkle spores over the surface, place the pot in a plastic bag, and seal the bag. The sealed bag will keep moisture constant on the soil surface. If the surface dries out, increase the moisture with spray from a misting bottle. Protect the spores from drying sunlight by placing the pot in a north-facing window.

The first leaves to show will be those of the prothallus, tiny, translucent green hearts. On the underside of the prothalli will grow both male and female sexual parts. The male sperm must have a film of water on the soil surface in order to swim over to the female parts. Once fertilization has occurred, the leaves of the fronds will begin to appear. When that occurs, the plants can be gradually acclimated to conditions outside the protective bag. Give the little ferns a week or two to harden off to the drier air, then transplant outdoors (if warm) or into separate pots. The whole process will take about six months, but, with care, you will produce *many* fern plants. Try to get access to a microscope once the prothalli appear. The bright tissues of these complex structures are very beautiful under magnification.

27

Mushroom Collecting

Despite the ominous reputation of deadly poisons and enchantments, the mushroom is always fascinating with its myriad forms and colors. There are some fungi (the plural of fungus) that *are* poisonous, but only to eat, not to touch. It is still best to instill a consciousness of potential danger by insisting that children wash their hands after handling any fungus. Please be equally careful in describing edible mushrooms to children. It would be wisest not to do so. The distinctions between mushrooms can be difficult for experts, and a thorough knowledge of all local species is a prerequisite to a single taste. The following are safe ways to enjoy learning about mushrooms.

SPORE PRINTS

Observing the colors and patterns of fallen spores is one of the ways mycologists (students of mushrooms) identify species. Each spore is a lightweight sphere that will fall straight down from the mushroom cap unless it is caught by a breeze and wafted away. Notice how the cap or shelf forms of mushrooms will orient themselves perpendicularly to gravity's pull so that the light spore will fall clear from their pores or gill slits. Were the mushroom to tilt, the spores would adhere to the underside of the cap and would never have a chance to grow fungus threads (mycelium) through the ground. A spore is not

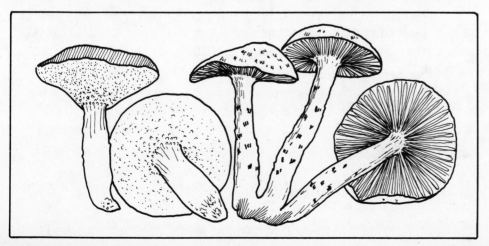

quite the same as a seed. It has no food supply or embryo plant inside it, but it is capable of starting the underground part of a fungus that may eventually send up a mushroom to produce more spores.

If the spores fall on a piece of paper, they will form a pattern that resembles the underside of the cap. Gilled mushrooms are visually the most interesting. As spores may be light *or* dark in color, you may have to do some guessing as to which color of paper will contrast with the spores. Putting both shades of paper underneath will eliminate the guesswork but is not as aesthetically pleasing.

Whichever way you choose, break off the stem so that the underside of the cap will rest close to the paper. For a very clear print, cover the cap with a cup or a bowl to keep the spores from blowing about. After several hours, or by the next morning, check the results.

As you handle the mushrooms, especially after making your print, check the surfaces and insides for insects, both adults and wormlike larvae. Tiny beetles and gnats (and their even tinier parasites) are specialists in mushroom habitation, completing their life cycles within the emergence and the senescence of a mushroom.

Use mushrooms as inspiration for models made of papier-mâché: flour, salt, and water "clay," or the kind of clay that may be "fired" at oven temperatures. Both will take poster-paint colors, which is half the fun of making models. However, you will need to refer to fresh mushrooms or to previously colored sketches by the time the models are ready to be painted.

For Discussion

In most ecosystems, the underground portions of fungi are the main agents for decay. The thin strands of fungi weave themselves into fallen leaves and dead wood, breaking apart and digesting the once-living plants, returning the nutrient elements to living organisms. This breakdown process makes it possible for other plants to use the nutrients for their growth. Some plants take shortcuts by tapping directly into the mycelium. The fungus, in its turn, takes fresh plant sugars from the invading leafy plant. In one study, young pine trees were found to grow faster in soil that contained fungus than in soil that lacked it. The fungus had made soil nutrients available to the young trees.

A mycelium is not especially hard to find in wooded areas. Check under or in well-rotted logs for delicate, weblike fibers. Notice the animals that take advantage of the soft, moist wood for protection and food. Fungi are prime agents in the creation and maintenance of most land-based ecosystems.

28

Haircap Moss Rain Meter

The large, furry haircap mosses, common on sandy, open soils, respond to moisture by expanding. This moss survives dry conditions of roadsides and gravel pits. When dry, its leaflike fronds are compressed closely about its stalk, which limits water loss from the tissues. When wet by rain, or even by saliva, the moss plant will spread out very quickly as special hinge cells take up moisture. If you have a large number of moss plants, each child can make a moss open by applying spit with fingers or by dipping the plants in water.

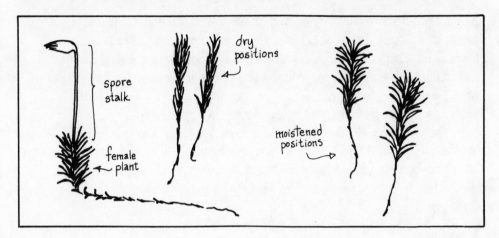

29

To Pick or Not to Pick a Flower

Most people who appreciate flowers began their interest as pickers of flowers. It is very hard not to wish to own the prettiness, even briefly. One way to point out to children the damage done by acquisitive actions is to present the group with two similar leafy stalks, such as lilac branches, and invite each person to pick "just one" of the leaves off one of the samples, then compare with the unpicked branch. There will seem to be little change after the first few pickings, but continued picking results in few leaves for the plant to use in food making and a bare-looking branch besides. Picking flowers does not interfere with photosynthesis; it may even save the plant the expense of flower and seed production. However, the loss of a flower does represent a decreased chance for seeds and, therefore, the future of a species.

For Discussion

Rather than "teaching" kids a rule about when to pick flowers, have them make up their own rule. Leave plenty of time for discussion of various possibilities. Be as democratic as possible when deciding on the final rule. Writing it down will give the rule greater weight for the children, even preschoolers. Be prepared for a stricter wording than you would have imposed. Most children have a strong sense of fairness when they can be the rulemakers.

30

Tiny Flowers on the Playground

A small plant with small leaves and small flowers takes little from its environment. In this way, many species of plants can specialize in habitats where there is little other competition. Cracks in sidewalks, dirt walkways and drives, and the edges of playgrounds are places to look for the tough and the tiny. These plants often compensate for their size by producing many blossoms, and these delicate miniatures can be quite beautiful.

The youngest children can be the most successful in this search. It helps to be close to the ground. Even without knowing their proper names, the children may be able to see that each species prefers certain conditions. Digging up the plants may reveal more interesting aspects. Digging will definitely give the diggers a respect for plants that are able to live in compacted soil.

As their plant parts tend to be thin or nonsucculent, these plants dry well when pressed in a book. Their small size makes them easy to use in decorative crafts, as well. After pressing for a week, glue the flowers onto an appropriate backing and cover the arrangement with transparent self-adhering plastic to decorate boxes, greeting cards, bookmarks, and so forth.

When fresh, the flowers can be enjoyed as small-scale bouquets for small-scale people, places, or toys. It's almost as much fun to think of household items to convert into tiny vases. Shampoo bottle tops, pen caps, thread spools, several soda straws stuck into a plasticine base are just a few possibilities.

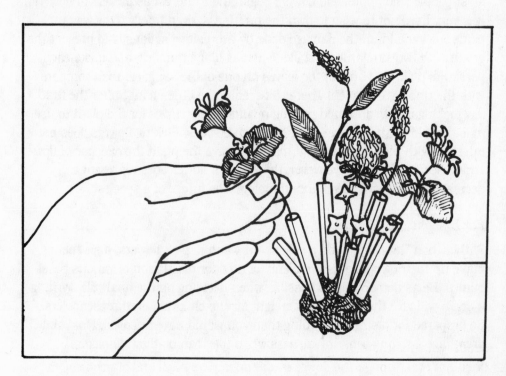

For Discussion

Spend some time observing the plants. What insects pollinate the tiny flowers? If you don't see any visitors, try putting bits of sticky clay of various sizes onto the flowers to find out how large a pollinator the stem could support.

31

Learning the Names for Parts of Flowers

For learners of any age, dissecting a number of different flowers will lead to a general understanding of the basic flower parts. Unless the leader is an expert, dissections will invariably lead to the discovery of unnameable parts. The parts of even the innocent-looking daffodil can be tricky to sort out. If a sense of fun and exploration is maintained, the children will come away from this first adventure eager to inspect other flowers with the tools of a new vocabulary. If they can find the stigma, the petals and the stamens, you will have given them the tools for understanding a great deal about flowers.

In general, flower parts are arranged in successive rings about the central focus of ovules, the unfertilized eggs. The ovules are protected within the ovary, which forms the base of the pistil. A stalk called the style arises from

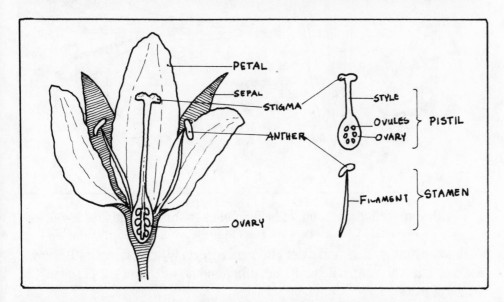

the ovary. At the tip of the style is the stigma, the sticky landing field for the pollen grains. The stigma has an adhesive surface that can be seen easily by dusting the pistil lightly with flour or pollen and noting the area where the powder sticks.

The next ring out is made up of stamens: slender filaments with polleny heads, called anthers. In some of the flowers you look at, the stamens may be dried up, even though the flower is still fresh. The pistil and stamens often ripen sequentially, a process that reduces the chance of self-pollination. The brightly colored structure surrounding the stamens and pistil is called the corolla, which may consist of petals but may also have colored sepals, which are petallike structures on the next ring out. Sepals are usually distinguished from petals in their function as protectors of the flower bud and are usually green. In day lilies, the sepals start green and then color up as the corolla opens and the inner petals are exposed. In roses, the sepals form a green case for the bud. They then fold back and out of the way as the petals spread and bloom. In both cases, they are clearly outside of the petal ring or rings.

In some flower families, there is a leaflike bud sheath called a bract, seen as the cellophanelike wrap at the base of a daffodil blossom. Bracts also serve as the brightly colored "petals" on the Christmas poinsettia, the white "petals" of the dogwood blossom, and the many green sheaths that protect the bud of a daisy, chrysanthemum, or dandelion.

If you feel nervous about facing all this diversity and possible confusion while trying to steer students through the nomenclature, look up the structures of the flowers you plan to use in a botany manual before the activity. If you are dissecting wild-growing flowers (only the very weedy ones, please), most wildflower handbooks will be able to help you.

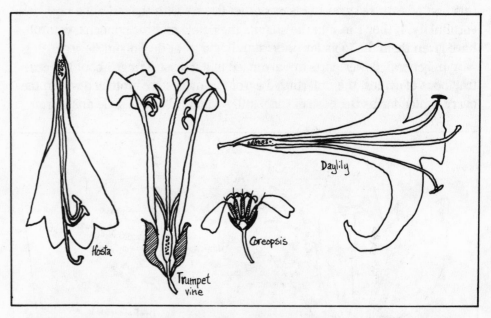

With young children, it might be helpful to go through the first few flowers as a group, to make sure that each child finds each part and is not overwhelmed. Third-graders and older children, especially adolescents, will enjoy working in small groups, helping each other and arguing over the possible names of parts.

For Discussion

■ Why do you think that the parts protecting the eggs are called "female" and the pollen-producing parts are called "male"?

■ Notice the ways that an ovary is often hidden away, the stigma is sturdy to support landing insects, and the stamens are usually frail but effective in spreading pollen dust. Do other flower parts have any other general attributes? Are they like other things that have similar functions? (Hidden Easter eggs, airport, paint brushes?)

■ Compare the parts of open flowers with unopened buds and spent blossoms. Describe and discuss the differences.

32

Sorting Flowers

If you have access to a variety of flowers from a generous gardener or florist, the project of making up a classification system will both stretch imaginations and enhance observational skills. Let the children make their own systems. Older children will enjoy working in small groups. If the children are not familiar with the vocabulary of flower identification, introduce the activity by sorting out a small group of flowers according to colors, sizes, or number of parts. Encourage each person or group to find unusual attributes, but caution them to keep their system based on things they can see. Although they may not use scientific terminology, the sorting process that they have experienced is the basis of all scientific study.

Reserve enough time for each group to explain its system of classification. This will be especially interesting if each group is using the same kinds of flowers. Use the words the children have created to distinguish attributes during other activities. Introduce scientific vocabulary where appropriate, but do not put down their new words by calling the botanical terms the "real" words.

33

Comparing Garden Flowers with Wild Relations

Lilies, daisies, carnations, and roses all have relatives in the field as well as the garden. If not the same species, they are often in the same family. Use pictures in garden catalogs to compare with pictures of wildflowers. Better yet, take garden specimens with you into the field, especially if you know where appropriate comparisons are. Pansies are a kind of violet, and peonies are a kind of buttercup. What kinds of changes in general have horticulturists cultivated in garden flowers?

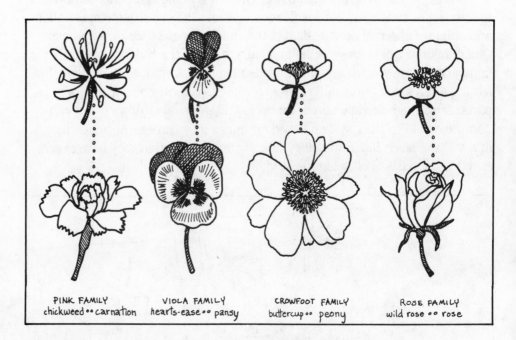

PINK FAMILY
chickweed •• carnation

VIOLA FAMILY
hearts-ease •• pansy

CROWFOOT FAMILY
buttercup •• peony

ROSE FAMILY
wild rose •• rose

34

Inventing Flowers

The information and experience gained by a dissecting activity can be incorporated into a number of activities. Individuals or groups can invent their own flowers, showing essential parts. Any art medium will do: paints, crayons, and colored papers can create three-dimensional effects. Consider animating the flowers by using materials that lend themselves to puppet making, such as stiff paper or cloth. The new flowers can then be put into motion, opening to the sun and pollinators, closing and going to seed. Make

up a story about the events in the life of a flower. It is very dramatic to see a
class full of flowers blooming and blowing, responding to imagined sun and
rain and darkness.

35

Calendar of Bloom

A calendar of bloom is a record of the days of blooming of local flora, both
herbaceous and woody plants. Keeping such a calendar is complicated for a
single individual, but a group can be organized to keep track of a number of
favorites. Each person can take responsibility for recording the days of blossom for several species. Children will gain a better sense of importance if the
information they gather is displayed in a clear and dramatic manner. A time
line, with a small portrait of each plant and units of time marked off to indicate the blooming time, will show the relative flowering periods. Encourage
discussions as to why some plants bloom longer or shorter times than others.
Introduce possible correlations of increased shade over the spring bloomers,
the influences of heat or cold, or the effects of the life cycles of insect pollinators. To keep the discussion lively, do not discourage any theories. With older
children, some of their theories may lend themselves to experiments the
group could try.

36

Nectar Guides—Road Signs for Pollinators

For all the pleasures we take from the beauty of flowers, it is for the pollinating insects that the plants hand out their colors. Each flower face is a set of
instructions as to how to enter the blossom and procure the promised nectar
or pollen. In general, a bright and showy surface or alluring scent catches
insects' attention from a distance. At closer range, the convergence of petals
to the center, sometimes accented with markings, shows visitors exactly
where to go for its reward.

Bull's-eyes and converging straight lines are common accent marking.
Sometimes the center is denoted by strongly contrasting colors. If the flower
you are looking at does not have distinctive markings, look for these possibilities: (1) a cluster of dots or a funnel-shaped petal may direct bees to the best
position, (2) narrow-petaled flowers may use the darker background to create
converging patterns, or (3) a bull's-eye that reflects only ultraviolet light may
be invisible to us but quite visible to the insect eye. Ultraviolet is a color to

bees, but we can see it only on photographs taken using ultraviolet-sensitive film.

The kind of animal that pollinates the plant relates to the color of the blossom. Red is a rare hue for native plants of northern zones. Bees are not sensitive to red (although the plants' odor may attract them). Hummingbirds are ardent seekers of red flowers, a fact to which the many red flowers of tropical plants attest. A typical hummingbird flower will also have no scent. Most

birds have poorly developed olfactory senses. The yellow, blue, and lavender flowers tend to attract bees, while white is likely to attract either flies (if the flower is fuzzy and musky smelling) or night fliers (if tubular and sweet smelling). These color rules are not strict guidelines; the deciding factor may be the odor. Bees go more for sweet smells, flies for the heavy musk smells. Some of the fly- and gnat-pollinated flowers go one step further toward resembling rotting fruit or meat and are colored a distinctive brown-red shade, which also tends to turn sunlight into heat. These are excellent ploys for a spring bloomer, such as the Eastern Skunk-cabbage, which can create a warm and smelly haven for potential pollinators.

One caution: Some garden species may have lost their nectar guides through horticultural selection. If they are of European origin, they are also more likely to attract honeybees, which are not native to North America, although they are often wild. Our native bees are the bumblebees and a large array of solitary bees, slightly smaller than the honeybee. Here at the Garden in the Woods, where native plants predominate, we see honeybees only in the early spring working the European Snowdrops by the cottage door. They are rare during the rest of the summer, leaving the rhododendrons, azaleas, and the prairie meadow flowers to be tended by the native bees, butterflies, flower flies, and wasps.

After inspecting a number of flowers for their nectar guide markings, create some floral advertising in imaginary flowers. Put out materials that will inspire the use of bright colors and large shapes. Use examples of native plants as starting points. Get the children thinking of modern advertising techniques for attracting attention and use those ideas for "superflowers."

37

Trick-a-Bee: Designing Flowers to Attract Bees

Flowers are signal devices to bring in potential pollinators. Not only are the blossoms elaborate come-hither advertisements, but their shapes often conform to the bodies of specific pollinators. A garden or field of flowers is a good location for testing which elements work best in attracting visitors. Choose a warm and sunny day to visit a garden or meadow. Late summer is best. Look at a variety of flowers and discuss which characteristics might be significant in attracting attention. Watch the insects approach and land on the flower and guess which view might be the most eye-catching. Make models of the flowers that are frequently visited. Try several designs to check for the variable of color, shape, and size. Make sure that the flower being tested is being visited that day, as many kinds of bees will work only one kind or color of flower during a particular day. And while flowers may also look open, if they are not producing nectar at the time you want to test them, the bees will not be interested.

When bees are busy collecting pollen and nectar, they are not likely to sting a snoopy human. Do not harass them, however. They will still defend themselves and sting if they feel cornered. Stinging will readily occur if you disturb the nest where the young bees are being raised. Keep your eyes open for ground nesters such as bumblebees and yellow jackets. The workers would be flying rapidly in and out of the nest area. Flight around a feeding site is more deliberate, with more hovering activity.

Incidentally, all this busywork to and from the flowers is mostly for the benefit of the growing larvae within the nests. Both native and honeybees use flower products for baby food. The pollen is altered slightly to make "bee bread" to feed the very young bees. The nectar is partially digested and condensed into honey, a long-lasting and high-octane food for young and old alike. Only honeybees make enough honey to make it possible for most to survive the winter. In native species only the new queens survive, and then, only as hibernators.

For Discussion

Keep records of which colors and designs are the most attractive to various insects. Use the data to create "superflowers" that a bee cannot pass up by exaggerating the attribute that seems the most alluring. Keep track of which pollinators visit the various models and try to figure out which information they recognize most easily.

38

Being Bees—Exploring Pollinating Flowers

Equipped with small twigs, cotton swabs, or a fingertip, children can imitate the actions of insects as they come in contact with the pollinating mechanisms of flowers. By following the lines of the converging lines or patterns (nectar guides), they can see for themselves the ways a blossom both receives incoming pollen and distributes new pollen dust. A snapdragon flower, for instance, must be pried open by a bee of a certain weight and strength. Closed petal doors restrict the smaller insects that might not brush against the anthers or stigma. By carefully moving a finger into the entrance of an azalea or rhododendron blossom, a child can observe the precise contact points of the curving stamens as they might distribute pollen on the back of a feeding bumblebee.

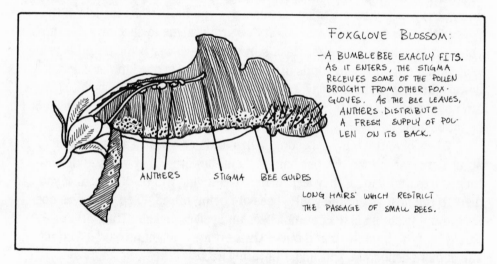

FOXGLOVE BLOSSOM:

- A BUMBLEBEE EXACTLY FITS. AS IT ENTERS, THE STIGMA RECEIVES SOME OF THE POLLEN BROUGHT FROM OTHER FOX-GLOVES. AS THE BEE LEAVES, ANTHERS DISTRIBUTE A FRESH SUPPLY OF POLLEN ON ITS BACK.

ANTHERS   STIGMA   BEE GUIDES

LONG HAIRS WHICH RESTRICT THE PASSAGE OF SMALL BEES.

Some of the pollination mechanisms will move dramatically when tripped by a bee's tongue as it probes for nectar. The stamens on the flowers of barberry or mahonia bushes will clap suddenly inward when tickled at their bases, like little mousetraps. The bee's tongue is effectively smeared with pollen. After several hours, the "trap" will slowly reset itself. When the curving stamens of a mountain laurel are pulled from their pockets by a fumbling bee foot, the released stamens spring upward, flinging strands of sticky pollen onto the insect or onto a hand held close to the flower. Garden peas and many plants in the legume family, such as lupines, also have spring-set stamens. The lower petals of the flower are fused together around the up-curved stamens, sheathing them with a narrow pocket. When a visiting bee steps onto the rim of the pocket, the stamen cluster springs upward, swabbing the bee's fur with pollen dust.

The flower has mechanisms for receiving pollen as well as distributing it. Look for the pistillate structures. Usually the stigma is held away from the stamens or extends beyond them, meeting the approaching bee while she is still likely to be covered with pollen from another plant. Pollen from another plant of the same species is of the greatest value to the flower. Another plant may have genes that represent slightly different characteristics, which may be encoded in the sex cells of the pollen grains. These different capabilities will be incorporated into the new seeds and therefore in the plants that issue from them. While the difference may be slight, it may be significant in giving the new plants a slight edge if the environment were to change.

For Discussion

At some point, especially with older children, it is important to address the question "Why bother with flowers?" The plant goes through all the fuss to produce these costly seed-making devices, but what does it get in return? If the children have really experienced the ways in which the pollen is spread about and the ways it is carefully received by the flower, they are ready to begin to appreciate the basis of sexual reproduction. In the production of pollen, the particular attributes of a given plant is spread abroad to new sites. In the seed, this is also the case. By *mixing* the plant's characteristics with those of another plant, that potential is given new capabilities. The species is maintained, with a potential for change that is essential to life on a changeful planet.

39

Weather Conditions and Pollinating Insects

As attractive as the flowers are to pollinating insects, cold or overcast days can limit the numbers of visitors and cut down on the fertilization of seeds. A heavy rain can wash out the pollen and nectar supply.

The effects of these conditions can be understood by comparing numbers of visits made during various conditions. The more patient, older child can be stationed at a particular flower for a set amount of time. It need not be a long time, but the times must be equal in length and at the same time of day. Some children might want to classify the kinds of visitors, as well. A good visual display of all the information gathered will reward the efforts and provide material for discussions and conclusions.

For Discussion

If the weather does affect the flower's opportunity to be pollinated, are there any ways that the plant balances out the odds? For instance, are there a series of buds that open over a number of days? Can one flower stay open for several days? Does the blossom stay closed during dark or wet days?

40

Violets—Discovering the Hidden, Self-Pollinating Flowers

As one of the first flowers of spring, the Common Blue Violet *(Viola papilionacea)* makes an eye-catching display of lavender-blue or white. If transplanted to a garden, the plants grow lushly and seedlings quickly take over all available space. New plants spring up everywhere. This profligate habit is due in large measure to the ability of all violets to produce special flowers that never open to insect pollinators but are held hidden under the leaves. These flowers are self-pollinating and produce many seeds. Without cross-pollination, there can be no mixing of genes to create offspring with new traits, so all the plants of self-pollinated seeds will display the same coloring and growth habits as that of the parent plants. (Blue Violets come in white with blue centers also.) This alternative way of flowers is called cleistogamy, which means, botanically, "fertilized in the bud." Once ripe, the stem that bears the seed-

Spring flower

Midsummer flower

pod grows out beyond the protective leaves. When mature, the pod dries out and pops open, broadcasting the seeds.

If violets are common in your area, this hidden method of producing seeds can easily be observed in a dug-up plant. Usually the cleistogamous seeds appear in late spring, after the showy blossoms are spent, but you may find some plants with both kinds of flowers. This is a good example of the high priority nature puts on cross-pollination. The showy blossoms always come first. The cleistogamous flowers are last, but their seeds ensure the continuance of that particular plant.

For Discussion

■ Cleistogamy is a backup to cross-fertilization. What might go wrong for the showy flowers to make it worthwhile for a plant to produce cleistogamous flowers?

■ If a disease hit the population of flowers, which seedlings would show the greatest resistance: those that are identical to their parent, or those that have possible new traits from a mixed genetic heritage?

41

Wind-pollinated Flowers

A bunch of pussy willows sold in the market in earliest spring is not usually thought of as a bouquet of flowers, but each silvery bud is a cluster of little flowers. On any stem, all the flowers will either be covered with pollen sacs (male) or with green, threadlike stigmas (female). Separate trees for separate sexes are typical of many wind-pollinated plants, but not all. What *is* typical is the obscurity of the flowers. We expect flowers to show bright petals as colorful advertisements for pollinators. But flowers such as pussy willows don't need animals to move their light pollen around (although bees do visit to eat the pollen). The plants instead produce abundant pollen and abundant seeds, both of which are carried away by the winds.

Most wind-pollinated flowers are catkins that look like swollen or elongated buds. Their overlapping scales give them a braided look. The typical pinecone is a female catkin. The woody scales are simplified leaves that protect the seeds until they ripen. Even an empty cone will still show slight

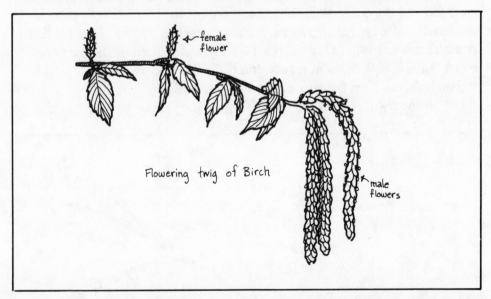

female flower

Flowering twig of Birch

male flowers

depressions on the inside of each scale where the seeds rested and grew. Male cones or catkins tend to be smaller. They are often visible on the ends of branches during the winter months. In spring they ripen and dangle, releasing clouds of pollen to the wind. When the pollen is gone, they wither and fall to the ground.

Most wind-pollinated trees bloom in the spring months (pussy-willow season) before the tree leaves are fully out. You may find insects visiting these early flowers. They will use the pollen for food. Look for wind-pollinated flowers on oaks, willows, birches, alders, poplars, beeches, hickories, most nut trees, and any pines, spruces, firs, or hemlocks.

For Discussion

Why do most wind-pollinated trees finish flowering before their leaves are fully expanded?

42

Grass Watching

We see grasses in lawns getting crew cuts so often that we forget what a lush and beautiful plant a grass can be. The leaves of grasses grow in a different way from the leaves and branches of other flowering plants. Grass leaves grow continually from their bases instead of from the tips of their branches or stalks as do the woody plants and most garden flowers. Pushing their leaves up from below not only helps the grasses survive persistent mowing in suburbia, but in their natural habitat of meadows and prairies, grasses can quickly grow back after grazing animals or fires have damaged the tops.

Narrow leaves and bushy growth are adaptations to a sunny habitat: a thinner leaf is less likely to get overheated and dried out by field-strength sunlight, and the dense stalks shade and protect the roots. Extensive root systems are also typical of grasses. For an impressive display of the holding power of a fine network of roots, plant a few grass or grain seeds in a paper cup 3/4 full of soil. Water at regular intervals. When leaves are six to eight inches high, cut open the cup and check the roots.

The flowers of grasses are not gaudy, but they are very lovely on close

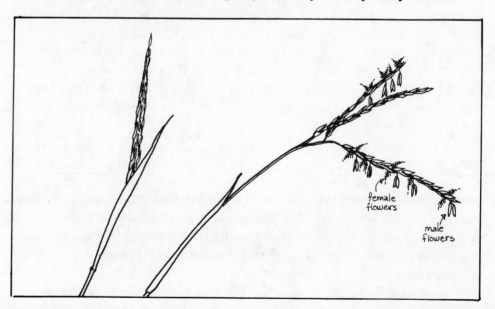

female
flowers

male
flowers

inspection. Children delight in the tickliness of the flower sprays, and a magnifying glass will reveal unexpected colors and forms. Collect a variety for close inspection. The male flowers will look like dangling, yellow hyphens, loosely attached and easily shaken by the wind. The female parts, where the seeds will ripen later in the summer, will show as tiny, twisted threads, tied tightly to the stem. The extended threads are stigmas. Usually male and female structures alternate along the stem. Later in the season, check the same grass plants for seeds. There are likely to be many tiny seeds that fall away at the touch of a hand.

Look for grasses in full flower in the early morning, or collect stalks that might flower the afternoon before you need them and place them in water as you would a bouquet of daisies. If kept away from breezes, the pollen on the morning blossoms is more likely to last into late spring.

For Discussion

Grasses are not primitive or less sophisticated than the more showy flowers of creature-pollinated plants. Grasses are very good at achieving cross-pollination using wind power. Ask children where they have seen grasses growing and try to find out what qualities make those places good grass habitats. Is there anything about the places that favors wind as a pollinating agent? Or could the location reveal clues about the ways seeds get planted?

43

The Giant Female Flower: Corn

Many of our foods, such as fresh corn, seem perplexing when viewed in terms of botanical generalities. If all seeds are the products of pollination, what does a corn "flower" look like? Corn is a very tall member of the grass family, most of which are pollinated by the wind. The flowers are not brightly petaled because the wind does not need bright colors to guide it. Most grasses have both male and female flowers on the same plant, placed one above the other, or alternating on the same strand. In general, the male flowers are above the female flowers. On the corn plant, the male, pollen-making flowers are at the top of the plant as large tassles of stamens that shed abundant pollen. The female flowers are lower down on the stalk and appear to be dense clusters of succulent strands of corn silk. Within the green sheaths of husks that cover the silk, each long strand ends at an ovule, an immature seed. If a pollen grain lands on the tip of a strand of silk (actually the stigma of a long pistil), the pollen will grow down and fertilize the ovule at its base, creating a seed. When the ripe corn is husked, the many silky pistils can be clearly seen. The fertile seeds are the plump kernels, and the ovules that are not fertilized remain small and undeveloped.

Before husking some fresh corn with a group of children, discuss what

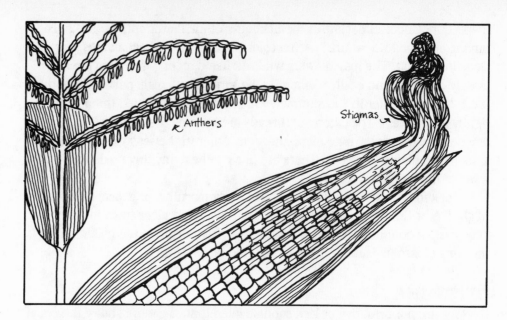

they think is inside. This will give you a sense of what they know and help establish a group vocabulary for describing the next step. Name each part and describe its function. The outer husks protect the silk and the seeds. The cob is the seed holder and the core through which passes the nourishment for the growing seed. Tell the kids that they are now fully qualified to answer the riddle: What food do you eat when you throw away the outside and eat the inside, then eat the outside and throw away the inside?

Try a few of the raw kernels for the taste of sugar. Do the undeveloped kernels taste as sweet? If you have some "older" corn, you will be able to taste the starchiness of the kernels as the sugar converts to starch in preparation for drying out and waiting for the next growing season.

44

Connecting Flowers with Seed Production

Our adult conviction that seeds come from flowers has developed over years of experience. For young children, the idea is very farfetched. The parts seem to have little to do with each other. At least, they *look* very different. For children, the time between flower and seed is very long. Finding the immature seeds, the ovules, in the bases of some varieties of flowers is a start toward understanding the connection. The flowers must be taken apart to discover the ovules within the protective ovary. Be prepared for rough and clumsy attempts at first. Eventually, after enough flowers have been investigated, the children will begin to recognize the patterns and anticipate the location of the seeds. Slice examples with a sharp knife so that the ovary can be observed in lateral slices as well as cross-sections. Don't use drawings to demonstrate, use real flowers.

Some examples of flowers with large ovules: all lilies, including related narcissus, amaryllis, and iris; roses, violets, poppies, and the daisy family. Daisies can be tricky, as the individual flowers are difficult to discern, but they can be dramatic. Sunflowers and dandelions have hundreds of seeds. If the children can watch for several consecutive days and have access to a dandelion patch, mark one of the flowers and observe its changes into fluffy, seed-filled seed head.

45

Self-Pollinating an Amaryllis to Produce Fertile Seeds

The amaryllis is one of the most dramatic of the flowering houseplants. The large bulbs can be purchased during the autumn from nurseries or mail-order catalogs and then potted in store-bought potting soil. Within two months, the huge flowers are blooming, much to the thrill of all who have watched the bulb's progress. The flower itself is a perfect model of flower parts: three colored sepals and three petals, six stamens and a single pistil, split at the tip into three parts. If there are several buds on one stalk, they

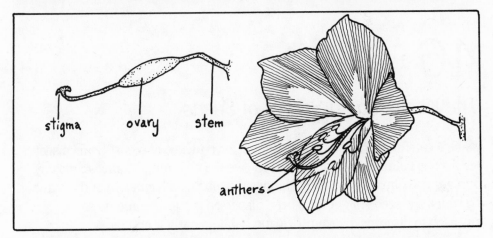

will bloom sequentially. If you watch carefully, you will see that the pistil tip (the stigma) and the stamen tips (the anthers) mature at slightly different times. The stigma splits open after the anthers are finished and dried up. When you see that the stigma is fully extended and exposed, transfer pollen from one of the freshly opened anthers from a younger blossom to the slightly woolly stigma. The pollen should adhere easily. This is *not* cross-pollination, as both the blossoms are on the same plant. You will need another flower on another bulb to achieve cross-pollination.

After the blossom wilts, leave the green swelling of the ovary on the stalk so that its slow enlargement can be observed. The stigma should still be visible, obviously a part of the seed-making system. You may wish to cut off the other blossoms as they wilt to funnel the plant's energy into the one fertilized ovary.

The full development of the seeds may take a month or more. Continue watering the soil whenever it dries out to the touch and supply the plant with extra nutrients by using liquid "plant food," as directed on the box. When the pod appears to be ripe (drying out), split it just a little and check on the color of the seeds. If the seeds are still white, they are not ripe and need to grow some more. If they are brown, you can collect them and plant them in potting soil. Cover them with about one-quarter inch of soil. Keep the soil moist to touch, not soggy. In several weeks, you can expect to have an abundance of tiny amarylli.

It will take several growing seasons before these little ones are big enough to bloom. If your climate is frost-free, the plants can be set in a shady garden and left to grow. Otherwise, they must be brought indoors during the winter frosts. On their own, or if allowed to dry out, the plants may die down. If the little bulbs under the soil seem firm, let the plants rest a few weeks, repot them in slightly more soil, and resume watering.

What color flowers should you expect? If the seeds were pollinated by the pollen from their own flower (self-fertilization), the seedlings will probably look like the parent plant. If the parent was itself a fancy variety, it may not be self-duplicating, and its offspring will reflect the flower-color characteristics of its parents. If you were able to cross-pollinate the seeds, you might get some of both parent colorings and some entirely new types.

46

The Kitchen as a Source of Seeds

Make a careful search of a home kitchen for evidence of seeds. Some discoveries may be botanical but not seeds, so give the children a chance to display and argue their cases for their findings. Encourage speculation of the manner by which the seeds might have been dispersed in nature, and do some research to discover the countries and habitats of their origins.

Try sprouting the seeds in moistened paper towel, kept in a plastic bag. Some of the tougher seeds, such as dates or peach pits, may take too long to keep young children's interest. Plant larger seeds in moist soil, label the pot, and cover with plastic to maintain an evenly moist environment. Many of our spices and fruits originate in tropical climates and are also likely to need warmth (over 70ºF) to trigger germination.

The following seeds are fairly reliable in their sprouting (unless too long on the shelf):

■ Any of the dried beans or uncooked whole grains

■ Coriander, dill, fennel (poppy seeds have been sterilized to prevent sprouting)

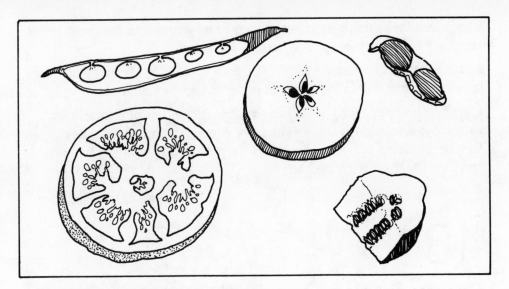

- Fruit pits such as apricot, plum, cherry
- Fruit seeds such as grapefruit, grape, orange, lemon, lime, apple
- Avocado pits
- Tomato seeds

For Discussion

Will the beans or peas from the fresh vegetables sprout? As the pods are opened, look carefully at the attachment and arrangement of the seeds. Try to get a sense of the ovary folding around the seeds that attach to alternating edges of the touching sides.

47

Activities Using Collections of Seeds

Maintaining a large assortment of seeds for individual activities will ensure material for many spontaneous projects. Consider some of the following activities and add some of your own:

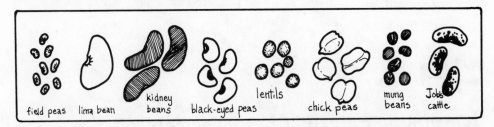

- Sort seeds according to size, color, dispersal methods, textures, or pairs of "opposites." Or make a spectrum of relative attributes: rough to smooth, light to dark, liked to disliked . . .

■ Use seeds for counting exercises or measuring games (one lima bean is equal to four lengthwise caraway seeds, and so forth).

■ Create pictures or mosaics with seeds. Pods and scales of cones lend themselves especially well to the design of flower shapes or sunbursts.

■ Write with seeds, gluing seeds into the shapes of letters or numbers, or write a name with glued seeds. White glue squeezed through a nozzle top can then be sprinkled with small seeds. Let the glue dry about fifteen minutes and then shake off the excess seeds.

48

Growing Pot Herbs from Seeds

Part of the pleasure of growing herb plants from seeds comes from being able to enjoy the distinct and pungent characteristics of herbs almost as soon as they show aboveground. This makes herbs good short-term garden plants for young or impatient gardeners. Most herb plants are fairly tolerant of crowded or dry conditions, a good survival skill for plants of first-time gardeners. Herb seeds can be obtained from flower catalogs or garden shops or can be found by raiding the kitchen herb collection. If the seed is whole and not too old, it will sprout and grow new plants. Most herbs have small seeds, so they should be buried not more than one-eighth inch in the soil. A seed as large as coriander seed can be put under one-fourth inch of soil. The soil can be sandy, but the addition of decayed leaves or peat will help the mixture hold water. After sowing the seeds and covering them, water the soil carefully (a flood will bring the seeds to the surface) then cover the soil with a piece of newspaper, cut to fit loosely inside the pot. The paper will hold moisture in the soil without encouraging rot, which a plastic cover tends to do. Peek under the paper every day or so to see if the soil is dry to touch (water again) or if the seeds have sprouted (remove the paper). Keep the soil evenly moist until the seedlings are large enough to be used as herbs. After two weeks of growing in a sunny, warm window, they should be ready to compare by tasting and smelling. Easy herbs to grow are: chives, cress, mints, basil, mustards, fennel, dill, and thyme. Try parsley, but be prepared for a slow-starting seed.

49

Which Are Fruits, Which Are Vegetables?

In botanical terms, fruit is the fleshy structure that surrounds seeds. We are more familiar with the term as it applies to apples, oranges, strawberries, grapes, and bananas, although bananas, as we know them, are a seedless variety that is propagated by cuttings. We are less prepared to call items such as tomatoes, eggplants, cucumbers, or squash "fruits," despite the seeds they carry. With the botanical definition in mind, ask the children to think of other vegetables that are really fruits. A good garden seed catalog, with color pictures of common produce, will help remind them and can be used to make a colorful collage to illustrate the concept. The pictures can be cut out and pasted on cards to be sorted in various ways.

A vegetable is botanically a more general category, referring to any vegetative part of a plant that is used for food. Vegetables may include leaves (cabbage, lettuce), stalks (celery, chard), roots (carrots, parsnips), underground stems (white potatoes, yams), flowers (cauliflower, broccoli), buds (brussel sprouts), or the seeds themselves minus the protective casings (corn, peanuts, beans).

A fruit, then, is the fleshy material that surrounds seeds. In a plant that is dependent on wind to distribute its seeds, a fruit is a means of getting an animal to swallow the seeds and then unwittingly deposit them at a new site.

50

Hidden Seeds in the Winter Earth

During winter or early spring, whenever the ground has thawed and can be dug, bring in a potful or so of surface soil from a number of different habitats and observe the growth of sprouting seeds. To make sure that none of the plants are sprouting from last summer's rootstalks, try to remove all the roots, stems, and bulbs that can be found. This kind of handling also gives children a good feel for the makeup of the soils from different areas. Be aware of different textures and proportions of sand, leaf mold, or clay. Make sure

that written identities of the habitats stay with the samples at all times. After the soil has been cleaned of obvious "nonseeds," spread it in wide-topped pots, water, and set them in the window or under lights. Keep records of numbers of seedlings and of varieties, and growth. Look for competitive tactics, as well. Some annual sprouts will quickly shade out competitors. At some point, you may want to pull out all visible seedlings and then watch to see if new sprouts appear, perhaps in response to the increased sunlight. Charles Darwin once did this experiment with mud from a pond edge and was amazed at the large number (several hundred) of seeds that sprouted.

51

Growing Peanuts from Seeds

As peanuts are such a staple in the diet of many children, the discovery that peanuts are the seeds of a plant can be very impressive to a youngster. Hand out roasted peanuts in their shells and have the children carefully open all their samples. Are all the peanuts alike? What is the texture of the coverings like? Is the relationship of a tough shell protecting the seed similar to any other plants (or animals) that they know? Carefully take apart the halves of the nut and find the embryo plant. Look for little winglike leaves and stubby root stalk. Taste as many different parts as possible. Any flavor differences?

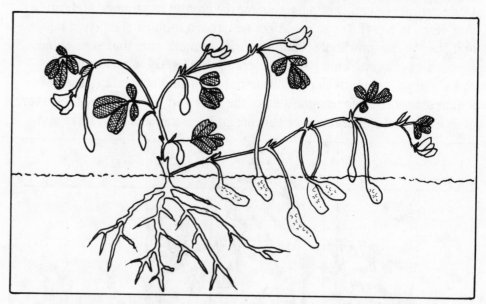

If you can get unroasted peanuts (health-food stores sometimes have them), you can compare the taste between cooked and raw. An uncooked peanut will grow a plant! Kids can plant peanuts and watch or record their growth. If time is limited, seeds can be started over a variety of times so that on the day of a demonstration, the children can see seedlings with roots and leaves at different stages of development.

To initiate growth of the seeds, use moist paper towels, kept warm (about 70°F). When the root is several inches long, the seedling can be carefully planted in an open-topped half-gallon milk carton, using a sandy mix of potting soil. Make sure there are enough drainage holes in the bottom of the carton so that standing water drains from the surface five minutes after watering.

Try to maintain the plant for about two months, until the seeds have formed. You will first see the flowers (very like a pea flower) bloom and fall off. They will self-pollinate, you don't need to help out. A green stalk (called a *peg*) will grow from the flower site down toward the soil. If the peg misses the soil in the pot, guide it into position. After it enters the soil, a peanut will form at the tip of the peg. The peanut plant plants its own seeds! After another month or so, when the plant begins to die and turn brown, pull out the roots, shake the soil from the plant, and check for peanuts. The more sunshine and balanced nutrients in the soil, the more peanuts you will have.

The peanut is in the bean family, many members of which live in association with nitrogen-fixing bacteria. Check on the peanut's roots for little light-colored nodules—the bacteria live there. This handy relationship means that the bacteria have a place to live and the bean plant has access to nitrogen, an essential component for plant growth. Beans and bacteria are the perfect example of a symbiotic relationship, in which two organisms live in need of and in benefit to the other.

52

Is It Ripe?

Seeds leave the parent plant only when they are ripe, that is, when they have enough of something. Ask the children what they imagine that something to be.

■ In most plants the seed must have sufficient food for its independent growth (oaks).

■ In some plants, the observable variable is the ripe covering, which becomes tasty or colorful enough to attract. The animals that eat the fruit will then transport the seed elsewhere (berries, fruit trees).

■ In some plants, the observable variable is a readiness to travel on their own (milkweed, maple keys, burrs).

If you have these kinds of seeds available, the kids will probably come up with these categories on their own. Be alert for some reasons that you haven't thought of before.

Encourage the children to act out ideas of "How will the seed get free to the earth?" to stimulate their creative problem solving. Fruits will have to be eaten out of their covering, burr seeds must be torn loose by a grooming animal. An acorn grows too large for its cap and pushes away from the branch.

After a seed-collecting expedition, sort seeds according to species and then for degree of ripeness. How can you tell? Does ripeness vary with the species? Is size always the clue?

53

Windborne Seeds

It is a pleasure to play with air-floating seeds, such as those of milkweeds, dandelions, lettuces, goatsbeard, thistle, or even goldenrods. Chasing after a buoyant seed, or attempting to blow it in a given direction in a race, gives

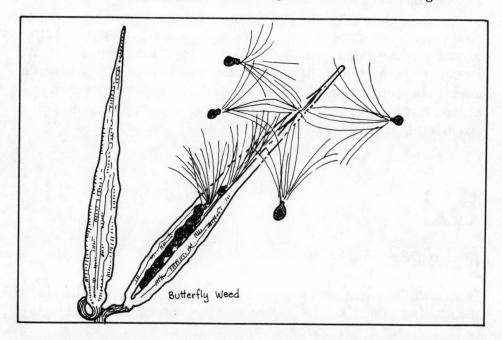

Butterfly Weed

children opportunities to use their bodies in unconscious and energetic ways. Before they get bored of chasing after the fluff, introduce one or more of these questions:

■ The seeds move in the same direction as the wind is blowing. Are there any other clues as to the direction of the wind?

■ Does the seed stay at the same level as it moves in the wind? What makes it rise higher or sink down?

■ Do all seeds released at the same time end up in the same place?

■ Can a grounded seed get back up into the air without being touched by a hand?

■ Test out more than one variety of airborne seed. Which ones move the farthest? The fastest?

If the inside space is appropriate for flying seeds, release some above a radiator or other source of radiant or forced-air heat and compare their journeys with other seeds of the same kind, released elsewhere in the room.

After some initial games and observations have spent some excited energies, use a magnifying lens (or an overhead projector!) to study the fine fibers of the fluff. What words describe its texture? How do the attributes described affect the seed's floating ability? Alter some of the plumes by delicate or major pruning and check their flying ability against an untampered plume or two. Does moisture or heat affect the plumes?

For Discussion

Use seed fluff patterns of movement to map the wind flow in a room or specific outdoor area. Do windborne seeds all tend to end up in one area? Describe the attributes of that area and look for plants with floating seeds in similar areas outdoors.

If you were to design a contraption to carry you away in the wind, what would it look like, based on the experiments with floating seeds? Where would you attach it to your body and what materials would you use? What might be some hazards of this as a means of transportation? Do seeds face the same problems? Draw pictures of the new floaters.

Similar games and questions can be played with spinning kinds of flying seeds, such as those of maples, ashes, lindens, catalpas, and trees of heaven. Try these seeds on another day, after playing with fluffy seeds, and encourage the children to use their floater knowledge to invent new games.

54

Seeds That Hitchhike

The simplest way to meet a hitchhiking seed is to go for a late summer/early fall ramble through any area where disturbed soil has been left to grow for at least a summer. An abandoned path, a roadside, and the edge of a playing field are also good places.

Examine the ramblers' socks, trouser legs, and so forth, for evidence of

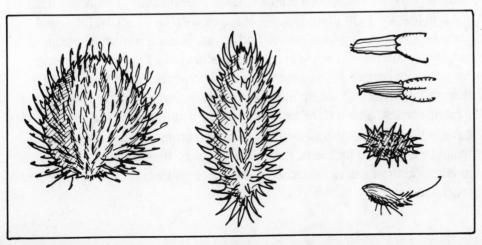

hitchhiking seeds. You may want to suggest that the children wear their scruffiest (most disposable) old high socks for the purpose, as the burrs are hard to get off clothing.

While extracting the burrs, keep track of the different ways each species uses to hold on. The triangular packets of the Tick-Trefoil have a short, dense plush of hairs. The Devil's Stick-tight is a narrow seed with barbed prongs on one end. The soft cone of the Burdock is covered with hooked spikes. (The story goes that a nearly bankrupt zipper company saved itself by inventing a new stick-to-itself fabric, called Velcro, based on the Burdock.)

If possible, find the plants that distributed the seeds. Where are the seeds located on the plants? Does the location help distribute the seeds? Notice what happens as the seed is extracted. Many seeds need to be "chewed" out of long fur to be released from the furry covering. Are the seeds themselves sticky, or slick and tiny? Many of our native animals that live near fields dig burrows for homes. The mounds of freshly dug earth on which they might rest at home and groom themselves is an ideal seedbed for these plants. The ground surrounding human homes is also a likely site for the plants, especially if a shaggy dog is in residence or shoes are cleaned outside. Check around back steps.

After exploring the attributes of various "stickers," it can be fun to create your own. Invite the children to imagine being unable to move by themselves and needing to catch hold of a moving vector (without its knowledge) that might be heading where the plant needs to go. Individual projects will produce greater variety, but group projects tend to raise the creativity level so that the results are more complicated and ingenious.

55

Seeds that Float in Water

Many of the plants that grow near water have seeds that can float. A river or lake can serve to transport a seed a considerable distance, eventually depositing it in just the right place to grow. Make a collection of seeds, including some of upland or dry-land plants, and have a floating contest, checking at reasonable intervals to see which ones are still afloat. Find out if the long-term floaters happen to live near water. What characteristics kept the best floaters from sinking? Set up a contest to invent devices for keeping seeds (perhaps beans) afloat. If the beans happen to sprout during the contest, refer back to the winning natural floaters. Because sprouting while they are afloat might be a problem for seeds, native floaters often have inhibiting seed coats or chemical timing devices that keep the seedlings from sprouting until they touch land.

56

Floating Cranberries

Cranberries are one of the best of the floating seeds. In their natural environments or in the bogs where they are grown commercially, a cranberry fruit floats away from its parent plant during autumn flooding. In nature, it can float to a habitat similar to its birthplace. Some of the attributes that make cranberries champion floaters are their tight but thin skins and their spongy flesh. These characteristics also make cranberries good bouncers. Fresh fruit that is appropriate for sale is routinely separated from damaged, mushy fruit by rolling the berries down an angled ramp. The good fruits bounce over a hurdle of established height. The damaged fruit don't bounce high enough to get over and fall to the side.

Pool the ingenuity of the children and set up a similar system to separate firm fruit from rejects. You might try turning the "bouncers" into a refreshing drink by squishing the fruit enough to break open the skins, mixing the mush with several times the amount of water, perhaps sweetening with honey, and serving.

For Discussion

Cranberries might remind the kids of other kinds of fruits. Try floating other candidates, as well, and try to find out why others will or won't float.

57

Berries and the Beasts

Locate a concentration of berry-producing shrubs that have not been planted by people and survey the immediate area for favorite bird perches such as fence tops, dead limbs, overhead wires, or protected night roosts. Birds will often defecate after arriving at or upon leaving such sites, effectively planting seeds of their preferred foods.

Experiment with finding out what attracts birds to seeds by collecting varieties of ripe fruit and offering them to birds. Choose a place where the feeding can be easily observed. Sometimes the first food a bird choses was the one most attractive from a distance, but another fruit may be preferred once the bird is up close. Try to get some idea of *proportion* of fruit eaten by comparing the amount (weight, number) of food before and after a day of feeding. As early morning and late afternoon are prime feeding times, try to include them in your survey.

Many native mammals that are active during the night are also berry eaters, so be sure to check before and after dark. Mammals also use their

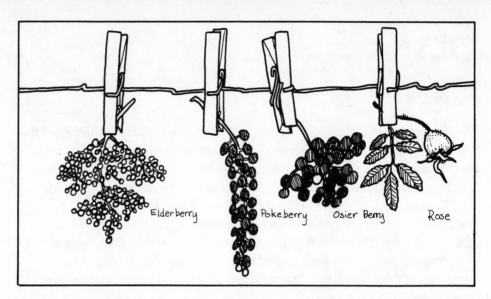

Elderberry   Pokeberry   Osier Berry   Rose

sense of smell, while birds are more visually oriented. Try hiding some of the fruit to test this idea.

For Discussion

■ Spend some time sorting the fruit and arranging various species in arrays of ripeness. What are the qualities of a "just ripe" fruit? Are there similarities among species? **Caution:** Do not taste *any* fruit unless you know it is edible *for people* in an uncooked state.

■ Mistletoe berries are effectively spread by sticking to the sites of a bird's bill until wiped off onto a branch where they will grow. Are there other berries that might be similarly messy "on purpose"?

58

Plants and Their Offspring— A Search for Origins

In this case, we're not searching for "roots" but trying to connect young plant forms with nearby mature plants through their similarities in leaf or flower. Trees and shrubs make the best subjects for this search as herbaceous plants are more likely to be short-lived.

Vacant lots are the easiest places to search for parents and offspring, as are old garden plots or untended yards. Look for an older stand of plants near an area that has been undisturbed for a year or more. Collect a variety of leaves and then "interview" older, larger trees and shrubs in the same area for similarities in leaf form or branching patterns. One distinction in woody plants is between those with opposite branching twig patterns and those with alternate branching. The game also works in the other direction: collecting examples from the older generation and searching for the younger one.

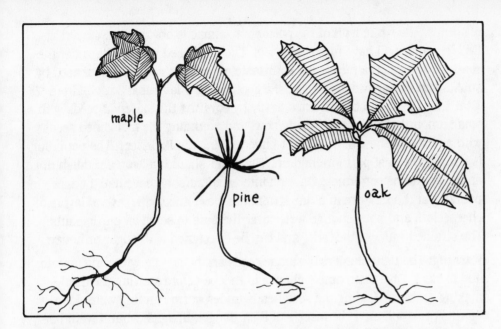

For Discussion

Try to bring out concepts such as relative numbers of parent to offspring, dispersal techniques, and observations on the relative success of the larger offspring. If no young are found, try to imagine what the area might have been like when the mature plant arrived as a seed. How has the habitat changed? Older trees, especially, can be great clues to the history of an area.

59

Sumac-ADE

In the activity on floating seeds, we described the preparation of a drink using crushed cranberries for flavoring. Another favorite wild foods treat, especially with young children, is a drink made from the red fruit of the various species

of sumac. The white fruit of the poisonous sumac is obviously excluded. Edible sumacs all have fuzzy red fruits. The hard seed can be felt under the coating of hairs. The pleasantly sour taste of the fruit can be experienced by sucking on one or two of the seeds, a good preview for impatient children. That sour flavor can be extracted easily by covering the loosened seeds with a small amount of water in a sturdy bowl and pounding the moistened seeds with a pestle of some sort. A blunt-ended rolling pin is perfect. The pounding is usually the best part for children. The water should take on a reddish tinge as the pounding continues. Once all interested persons have had a chance to pound, put the mixture in a fine strainer or sieve lined with several layers of cheesecloth and pour enough water over the pulp to serve all participants. The strained water is the "ade" and can be sweetened with sugar or honey.

Warning: The tight clusters of sumac seeds are homes for wintering insects and spiders. Give them ample chance to escape before starting the drink-making process. Holding the seed heads under warm running water for a minute or so will alert the insects of their danger. Imagine what it would be like on a wintery day in a dense cluster of furry, red seeds.

60

Acorn Play

Oaks with pointy-tipped leaves spend two years growing their acorns, and the shells are lined with a soft fuzzy coat. Oaks with rounded leaves with no points take only one season to grow their acorns, and the shells have no furry lining. The volume of either acorn crop will vary from year to year. Usually there is a cycle of a very good year followed by several seasons of relatively

reduced production. If you notice that acorns are abundant, here are some activities to try.

Encourage the children to think squirrely thoughts and collect as many acorns as they can. Then gather the kids and their acorns in one place and discuss the experience. Was it hard to find acorns at first? Did they find they got better at recognizing the shapes and colors as they collected? Considering the area that they had for searching, where did they find the most acorns? Were acorns found only under the oak trees? Most plants have some way of transporting their seeds to new places away from the shade of the parent. (Recall milkweeds and so forth.) How might an acorn travel? (Squirrels and jays have been found to be major movers.)

What ways could the acorns be sorted into categories? Do the differences in size or color indicate a difference in oak species? (Look at scaliness of the acorn cap, the furriness of the inner shell.) Is there evidence that the acorns are used for food by animals or insects? Try cracking open a few. The grubs of acorn beetles may still be inside. If they have already exited, you will see small, round holes. If you put a number of acorns in a bag or jar, some adult beetles will emerge later and can be observed.

What kinds of uses can you find for acorns? Play with the acorns, exploring their attributes for smoothness, hardness, roundness, and so forth. What other objects have those attributes? In the past humans depended on acorns as food. The acorns were shelled and buried in a stream or soaked in water until the bitter-tasting tannin was removed, then the nutmeats were ground up and baked, often with cornmeal as breads.

Try soaking a number of acorns, after cracking the nuts with a hammer first to hasten the process. A pink-brown coloring will stain the water. This is tannin, a chemical created by the plant to discourage insect, bird, and mammal predation. (Although, when acorns are ripe, the predators that store the acorns in the ground serve an essential role as carriers and planters.) Tannin is used by both Native American and Western cultures to soften and preserve animal skins as leather. It is also a dye agent. Simmer some white *wool* yarn (not acrylic) in tannin to color it a warm brown.

Plant some acorns in pots and watch for sprouting. If sprouting does not occur within two weeks, the acorns may require a month in the freezer. The cold helps to break an inhibition that keeps them from sprouting.

WHISTLES

Another use of acorns, or rather, their caps, is as an aid to producing a loud whistle. Bend both thumbs symmetrically over the upturned cap, and blow evenly across the shallow cavity that is formed in the cap. Experiment until you get a whistle. By flexing the thumbs, a variable pitch can be obtained. Kids become very fond of this loud noise, so don't introduce the skill unless you are willing to put up with the racket. Better yet, put the skill to use in a game of hide-and-seek or as a means of communicating coded information over a distance. (See illustration on next page.)

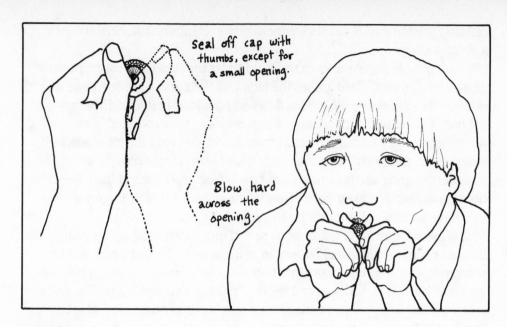

seal off cap with thumbs, except for a small opening.

Blow hard across the opening.

61

Pinecone Rain Meter

The tough, woody scales on the cones of many pines, spruces, firs, and hemlocks will respond to dry conditions by opening out, spreading their scales so that the seeds beneath the scales are released to the winds. When the cones are wet by rain (or by dunking them in water) the scales close, which protects the seeds from being washed down under the shade of the parent tree. The scales will respond to moisture by closing fairly quickly, but will take a day or so to dry out again. Have a number of cones on hand to experiment with various moist conditions. If the cones are still green and tightly closed, the scales won't react to moisture. The seeds inside are probably still growing. It is interesting to look in the cones for the seeds. The seeds have delicate "wings" of stiff tissues. Even if there are no seeds, you will be able to see the impression of the seedbed, usually two per scale, at the base of the scale.

62

Making Raisins

While we may have been told any number of times that raisins are dried grapes, going through the process finally makes the connection real. In order to get an idea of how much water is lost from the grapes to make them into raisins, weigh the grapes now, once they have been picked from their stems and sorted, and again immediately after they are dried. Don't bother to keep very small or bruised grapes.

Rinse the grapes under cold water to clean them. If you believe pesticides were used at any point in their growing, slosh them in warm water with a small amount of liquid soap, then rinse thoroughly with fresh water.

Next, drop the grapes into boiling water, then pour the water off the grapes immediately. This breaks the skins just enough to hasten the drying of the pulp and helps to sterilize the skins against molds that might start growing during the drying process.

Spread the grapes evenly, one layer only, on a cookie sheet or shallow pan and set them in an oven turned to about 160°F, or "low." The drying will take from two to six hours, depending on the size of the grapes and the relative humidity in the air.

As you have to weigh the grapes again, don't taste them to see if they are "done," unless you have put some grapes aside just for tasting. Then weigh and calculate the proportions of water to grapes. Do the grapes, or rather, raisins, taste sweeter when dried? Which condition resists decay longer if the fruit is left out of the refrigerator? Does it make sense that making raisins and other dried fruits such as prunes was originally an early method of preserving food?

63

Pumpkins: Beyond the Jack-O'-Lantern

Pumpkins are the largest members of the squash family. Make any investigation of its insides appropriately dramatic. When the insides are scooped out from the top for the traditional lantern, the cross-section pattern of seed arrangement is missed. Try cutting a pumpkin in half (either way) for a clearer view of the insides as well as to produce two study specimens. Once kids get involved in the cleaning, the process is satisfyingly messy. Lots of paper on the floor and tables will allow for thorough investigations.

At suitable times during the cleaning activity, notice and encourage speculation:

1. What do the strands connecting the seed to the inside of the pumpkin wall do?

2. Where are the seeds clustered?

3. Is there a design or pattern to the cross-section?

4. Are all the seeds the same size?

5. Will pumpkin plants grow from both sizes of seed?

6. What do the differences between the top (stem) and the bottom (where the flower attached) indicate? Is that arrangement similar to other fruits? (Have an apple ready for comparison.)

7. What does a pumpkin look like when it is flowering? If the children don't know, and you haven't a picture, have the children make up an appropriate flower.

8. What would it be like to walk into a pumpkin the size of a room?

9. What ways could we use different parts of the pumpkin around the house?

After the seeds are separated from the stringy parts, rinse them in cold water (in a sieve is easiest), blot them with a towel, and cook them, either by roasting them in a 300°F oven on a cookie sheet (the easiest way) or over medium heat in a frying pan with a little oil. For both methods, shove the seeds around every five minutes or so. After fifteen to twenty minutes, they should be brown and crisp. Sprinkle some salt or soy sauce on them to make them more "snacky."

While the cooking is going on, investigate the insides of other squash family members. Look for similarities in seed shape and arrangements of seeds in cucumbers (or pickles), yellow squash, zucchini, winter squash (seeds are also good cooked), loofah sponges, melons, and gourds.

64

Ripening Fruit Using Ethylene Gas

A piece of cut apple releases small amounts of ethylene gas—not enough for us to detect, although the apple smells pleasant, but enough to cause unripened fruit to ripen. Try an experiment with a couple of bananas. Choose two green bananas from the same bunch or "hand." Place both bananas in plastic bags, just to keep things even, but put a slice of apple in one of the bags. Close both bags. Observe both bananas over a period of days to see which one ripens first. Ethylene gas also hastens the ripening of tomatoes and whole apples. Perhaps ethylene is the culprit responsible for the adage about one "bad" apple spoiling the whole bunch.

There is some evidence that trees respond to ethylene, as well. An experiment indicated that trees under attack by leaf-eating caterpillars were able to defend themselves by producing a bitter-tasting substance in some of their leaves. In the process of crawling around to find the good-tasting leaves, the caterpillars exposed themselves to predators and took longer to grow into moths. Both results diminished the damage to the tree. There was further evidence that adjacent trees, not under attack themselves, responded in the same way. Ethylene gas from the torn leaves seemed to be the link.

65

A Coat for a Seed

The tough and sometimes colorful covering of the seed is essential to its survival. The seed coat protects the embryo plant and its food supply from invasive fungus, rot, and insects. Without it, the embryo would be exposed and vulnerable. However, the seed coat must be weak in one spot. It must have

some small opening so that when conditions are right, water can reach the
tiny root and initiate the first stages of growth.

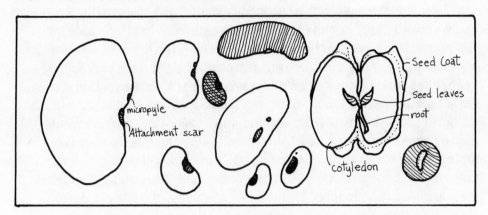

In a bean seed, that spot is called the micropyle. It is barely visible to
one side of the scar where the seed was attached to its pod. When beans have
been soaking in water for a half hour or so, the seed coat surrounding the
micropyle will look puckered and wrinkled where the water is starting to seep
in under the coat. After an overnight soaking, the entire coat will have a wrinkled look and will be softer to touch and easier to remove.

The location of the micropyle can be observed on dried beans by drop-
ping the beans into very hot, just-boiled water. The air under the seed coat
will be expanded by the sudden heat and can be seen escaping through the
micropyle as a line of small bubbles rising to the surface. The effect is short-
lived, so have a number of beans available to make sure everyone sees the
reaction and the location of the hole.

The micropyle is also the "escape hatch" for the root, the first part of the
baby plant to push through the protective coat.

Bean seeds are by far the easiest seeds to explore as they have large,
clear parts. They are also easiest to find. They are abundant in grocery stores
and well represented in common flower and vegetable seed packets. Scarlet
runner beans have especially gaudy coats.

For Discussion

After looking at a number of seed coats, make a list of all the advantages of a
coat for a baby plant. Discuss what kinds of problems the embryo would
encounter if it had no coat. What other "coats" do the children know about
that protect organisms from similar dangers?

66

Sprouts: Meeting Them and Eating Them

Sprouted seeds of mung beans are often available in grocery stores or salad
bars. Make your own by soaking a handful of mung beans overnight and then

draining the water. Keep the beans moist for another day, then try the following explorations.

Once the outer, protective coating is removed, the bean seed will divide easily into two parts, or halves. Nestled between these two thick pads, or cotyledons, is the small embryo plant, complete with a tiny root and two tiny leaves. The cotyledons are filled with carbohydrates (starches and sugars) to nourish the young plant in its first days of growth. When its roots and leaves are sufficiently developed, it will be able to make its own food using sunlight energy to take apart carbon dioxide and water molecules. These elements will then be used to make more carbohydrates. (Notice that the word *carbohydrate* is made of two parts. *Carbo* refers to carbon, and *hydrate* is a scientific word used to denote compounds containing water.) The parent plant takes the sugar that was produced in the leaves and removes one molecule of water; then it combines many of these new units to make large starch molecules that do not disolve and can be stored for a long time. This starch is stored in the cotyledons. When water is added to the dry seeds, enzymes in the cotyledons initiate processes that digest the big starch molecules, changing them to sugar and making the food available to the embryo plant.

For eaters of sprouts, plant sugar means easy energy. Sprouting also increases other food values. Vitamin C is greatly increased in the first days of the sprouting process. A one-pound package of dried beans increases to six to eight pounds of sprouts.

Sprouts can be eaten in salads, ground up into drinks, or stirred into soups. In general, cooking sprouts destroys much of the vitamin C, so for the eater, the fresher, the better. Many kinds of seeds can be used as sprouted food.

Warning: Don't use any seeds packaged for garden use unless you are sure that they have not been chemically treated with insecticides or mercury compounds. The package will usually tell you only if they have *not* been treated, not if they have. Safe seeds can always be found at Asian food stores or at health-food stores. Suggested are the following: alfalfa, mung beans, soybeans, fenugreek, peas, sesame, wheat, corn, or any whole grain.

TO GROW YOUR OWN

A wide-mouthed jar, a covered plate, and a clay pot are all possible sprouting containers. You must be able to get fresh water in and then drain it out easily. The seeds will expand and extend greatly, so a large opening will be helpful in removing the sprouts.

Soak the seeds overnight and drain in the morning. After that, rinse the sprouts twice a day to provide fresh water and wash away possible wastes. Sprouting goes faster in warmth, slows down in cold. The sprouts may be kept in the dark or in sun. Eat them when they taste best to you, perhaps in two to four days. Keep them moist but well drained to prevent souring. Refrigerate the sprouts to stop the growth.

POSSIBLE INVESTIGATIONS

■ Taste the various parts of the sprouts for their comparative sugar content.

■ Mark on the leaves and young root in small, even units, using a fine, water-proof pen. Let the sprouts continue to grow and note the areas of expansion. What hand gestures would express leaf growth and how would they be different from the gestures describing root growth?

■ Use tincture of iodine to test for starch on various parts and on the same parts at different growth stages. **Warning: Iodine is Poisonous. No tasting!** Use the iodine applicator to touch a drop of iodine onto a piece of the seed. If starch concentrations are high, the drop will turn blackish. The reaction will be clearer if the tissue has been broken up enough to release the starch from the cells.

■ Discuss with the children what they would feel like if they were sprouting seeds. What words express the feelings? Would the feelings change as the sprouts grow? What are dangers to sprouts? How would it feel to begin to dry up? to be watered? to open leaves to the sunshine?

67

Sprouting Gestures

Different seeds sprout in different ways. The first leaf of a corn rises straight as a shaft from the ground. A bean sprout looks lazier: It lifts its heavy cotyledons aloft by means of a crooked elbow. A pine seedling pops out like an umbrella. Observe first gestures out of the seed coat of a number of seedlings. What do the typical gestures look like when acted out by a person? Or play a charade-type game in which a team must guess the species of a seed from the charade performance of a member of the other team. Keeping a chart or a

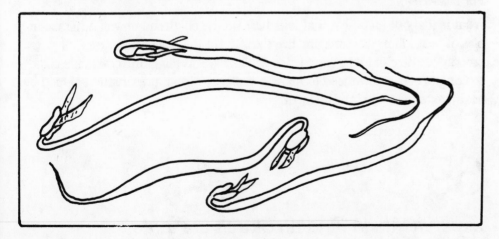

bulletin board for drawings of the sprouting activities will help build a gesture vocabulary.

68

Cooked Beans/Live Beans

When introduced to the experience of growing bean plants from soup beans, young children might wonder if the beans eaten in soups might start growing in their stomachs. Find out from your children why they think this does not happen. Introduce the following exercise as a way of demonstrating how cooking stops the potential for growth in the little plant within the seed, making the food available for our bodies. Point out that the grinding action of our teeth on fresh sprouts has the same result.

Prepare for the experiment by soaking overnight a dozen beans per person or group. Divide the dozen beans between two zip-lock bags so that each bag contains two beans to look at after cooking plus another four or so beans for a growth experiment. Put a small amount of water in each bag. Try to get all the air out of the bags, then seal them. Set one of the bags in water that is just under boiling temperature for at least half an hour. Cool the cooked beans under cold water before opening the bag. Mark the bag containing cooked beans clearly.

Compare the cooked beans and the soaked, uncooked beans for as many attributes as the children can discover, including taste if you are using beans purchased in the supermarket's soup section. Place two examples of seeds from both treatments in equal-size flower pots filled with the same kind of potting soil. Try to water the pots equally, so that the only different aspect is the cooking of the one set.

While the growth experiment is going on, leave a few of the beans in respective bags and observe their changes. The cooked beans should rot and smell and the soaked beans should begin to sprout. If no beans sprout, you may have inadvertently picked up beans that have been on the shelf too long.

For Discussion

What is it about high heat that kills life? Do the children know of other examples of heat killing or damaging live things? Let the discussion range as broadly as the children's experience will allow. You might introduce ideas of cooking to break down food to make it easier to digest, preservation of food by heat (pasteurization), and so forth.

69

Experiments in Finding Starch in Plants

Tincture of iodine will color starch molecules a dark brown or black color. Materials lacking starch will color brown-red. This sensitivity to starch makes iodine an indicator substance, a chemical Sherlock Holmes, and children can experiment with a variety of foodstuffs and plants materials to discover the starch-containing parts.

The green in a plant's leaf indicates that photosynthesis is possible in those tissues. Sunlight on a leaf powers the process of making sugar from other materials. The sugar travels throughout the plant as food for growth. Starches are altered sugars, compacted by the plant and stored in roots, stems, and leaves for later use. Starches are stored in seeds, as well, and are reconverted into sugars as the seedling plant grows and draws on its sugar "account." We usually cook food starches to begin the breakdown of the dense starch molecules and make the food easier to digest.

With this background, test for the answers to these questions:

■ Which would show a stronger starch stain, uncooked or cooked potatoes? Bread flour or a slice of baked bread?

■ Many garden vegetables have been cultivated for their sweetness. Which shows the greater starch content, a store-bought carrot or the first-year root of the Queen Anne's-lace? (Both plants are the same species.)

■ Do different parts of a sprouted bean stain differently? Does mashing the part make a difference in the stain?

■ Does the starch content in a bean cotyledon change as the seedling grows?

■ Compare the starch stain on a unchewed Saltine and a piece of well-chewed Saltine (or bread). **Remember, iodine is a poison. Do not taste!**

■ Touch a drop of iodine onto the substance to be tested. If starch is present, the drop will turn a dark brown or black color. (The iodine solution can be diluted with up to four parts water to one part iodine and still be effective in detecting starch.)

Health-food stores are full of items to test. You can often buy small amounts of a great variety of uncooked seeds and grains. To get you started with items you might have around the home, check this beginning list:

Dried and sprouted beans

Corn, corn chips, corn starch

Potatoes, mashed potatoes, potato chips

Tapioca

Carrots

Celery

Flour, bread, crackers, doughnuts

Peanuts, peanut butter

Various Parts of Plants: Names and Functions

70

Water Movement Through Plants: The Veins

By definition, a higher life form organizes its cells into particular functions; in the case of plants, water and nutrients are transported by specialized cells from the root to the leaves. Children may be familiar with the system of veins that carry blood through their body but not know of veins within a plant.

Pale-colored plant parts such as celery or white flowers are best suited for this experiment. Place the cut end of the leaf, stalk, or flower in diluted ink or water colored with food coloring. The color can then be observed to spread through the veins. There is usually some noticeable effect in an hour. What happens if the plant is left in the coloring for a week?

Celery is an excellent subject for vein study. The coloring can be seen as it progresses through the plant. Red or blue tints make the best contrast. Try two pieces together, one with leaves on, one with the leaves cut off. Cut crosswise, the positions of the veins that are transporting the color can be easily seen. Any celery eater knows that the veins are tough enough to provide some structural support, as well.

The project above combines well with the experiment to remove leaf tissue to observe the veins. Try coloring the veins by this method prior to removing tissue to provide greater contrast.

For Discussion

Celery with leaves should color up faster than the leafless stalk. Encourage guesses on how the leaves might be involved in moving water through a plant.

71

Making Leaf Skeletons

To see the plumbing that runs through every leaf, you must remove the green tissue parts. This process works best with tough kinds of leaves, such as the larger kinds of houseplants (see illustration of rubber plant), evergreen broadleaves (southern magnolias and evergreen oaks), or larger oak, hickory, or beech leaves. If these are not available, try other leaves in late summer or early fall, when the leaves are a little tougher than in springtime.

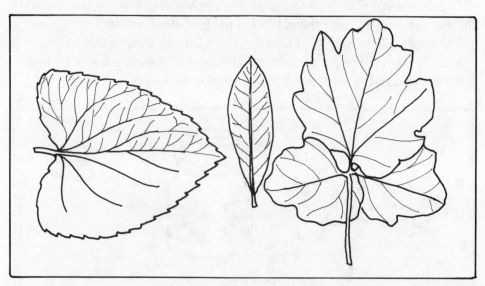

Prepare a solution of washing soda and water by dissolving about three tablespoons of washing soda (from the supermarket) and a quart or so of boiling water. **WARNING: Washing soda is poisonous. No tasting!** After the soda has dissolved and the water is only simmering (not boiling), add the leaves. Keep the water simmering for about a half an hour or until the green parts start to feel soft and slimy. Try a number of leaves so you can experiment with the best length of time for your plant. Take out a leaf and gently rub away the green tissue with an old, soft toothbrush. This part takes gentle concentration. Try to get one leaf cleaned without breaking any of the veins. The leaves can be dried flat in three days by pressing them between a couple of layers of absorbent paper weighted by a book. The leaf veins can be preserved for further study between two layers of self-adhering clear plastic.

For Discussion

■ Some people think that veins look like city maps. Have some aerial views or road maps for the children to compare. What might be the reasons for the similarities?

■ Look for the veins in people's bodies, especially the wrist area. Do they serve any of the same functions as leaf veins? What are the differences?

72

How the Leaves Help Pull Water Through the Plant

The bottom surface of all leaves has many tiny portholes, too tiny for the unaided eye to see. These holes are very like mouths in that they can open and shut to control the passage of water and gases. They are called stomata (one is a stoma), which is Greek for "mouths." As water evaporates from an open stoma, more water is pulled up along a vein that stretches all the way down to a root. This root water also carries dissolved minerals that the plant uses to build cells at growing sites. Some of the water will be used to carry sugars from the leaves to the growth or storage sites. Most of it will pass through the stomata into the air.

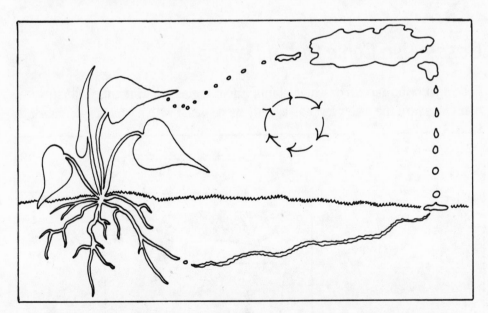

If leaves are left on celery or carrots and the plants are left unrefrigerated, the firm plant parts will soon become flexible and soft. Leaving the plants in sunlight will hasten the process. Keeping them in a refrigerator will slow it down. A good comparison can be made between leaved and leafless plants of the same size that are exposed to warm air or sunlight for the same amount of time. Ask the children why they think the plants need to keep stiff instead of flexible.

Another investigation involves placing leaved and leafless celery stalks of the same size in glasses with equal amounts of water and noting the difference in water loss from the glasses over a period of time. Place another pair of samples in the sun for a speedier loss of water. Is the proportion of water loss the same for the two situations? This movement of water through the plant and out of the leaf is called transpiration. Can transpiration be limited by coating the underside of the leaf with a water-resistant substance such as petroleum jelly? What effect does coating the top side have? Both sides?

The transpiration of any growing plant can be seen by placing a plastic bag securely around several leaves on the end of a branch or completely around a potted plant. After a day or so (especially if the experiment is in the sunlight), water can be seen in the bottom of the bag or condensed on the inside.

For Discussion

■ What effect does perspiration have for people? Do plants have similar needs of cooling or cleaning themselves of wastes?

■ What are some other examples of water vapor becoming visible as condensation?

73

Gases Also Come Out of Leaves

Collect some aquarium or water plants under a glass in water. (See illustration.) Measure the water level as exactly as possible with a wax crayon mark,

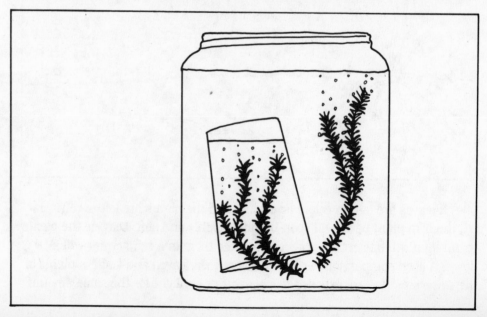

or make sure that there is no air in the glass so that when air collects at the top, it will be only from the plant. If possible, place the experiment in a sunny location. Observe the accumulation of gas released from the plants. Bubbles may be observed rising to the surface if the water is warm and the sun is striking the plant. Keep track of each day's increase in air in the jar and correlate the amounts with the approximate number of hours of sunshine for each day. Would the experiment work the same with a light bulb as a light source? Try increasing the light with aluminum foil as a reflector. Or use a special plant light.

74

A Leaf is Full of Water

Not only the transport system of a plant, water also makes up much of the mass of a plant, giving substance and uprightness to tissues, especially leaf tissues. A plant deprived of water quickly shows its loss by wilting. Miming this look even feels sad. How much the leaf depends on water in terms of weight can be measured easily by weighing a succulent leaf such as lettuce in its fresh state and then after drying. The difference can be dramatized by measuring out some water to equal that lost by the leaf.

When the figures are placed in a proportional relationship (the weight of the "before" under the weight of the "after"), the figures express a fraction. For young children, round it down to at least fifths (the digits on one hand) to help them understand the relationships. For instance, three ounces of fresh leaves might dry down to one ounce of crispy dry leaves. A whole leaf could then be seen as two-thirds water and one-third dry cellular structure and carbohydrates. If the dried material were burned (which would use up the carbohydrates), the remaining ashes would approximate the amount of "stuff" contributed by the nutrients from the soil. The fractional expression also gives you a means of comparing other plants that are measured and dried in a similar manner. Ask the children to guess beforehand which leaves they think are the "wateriest." Aquatic plants will tend to have the least cellular structure proportionally, and plants of dry habitats and evergreens will have the most. (Not counting the cacti, which can store large amounts of water.)

For Discussion

Q. How would a dense cellular structure help contain the water for plants in a dry climate?

A. Desert plants have extensive root systems to gather water and nutrients dissolved in the water. Aquatic plants have few roots. They absorb most of their water and nutrients directly from the water about them. What predictions could be made about the thickness of the skin on a water plant?

75

A Wilting Contest

After experimenting with the various situations under which a cut plant or stalk with leaves wilts, set up a wilting contest. Use the same kind of plant with the same amount of leaves for each contestant. Celery is a good choice because the flexibility of the stalk helps to establish "degrees" of wilting. Have contestants create situations that they think will cause the plant to wilt. Make sure that everyone has an equal grasp of the possibilities by discussing the effects of hot/cold, light/dark, and dry/humid variables. After a designated amount of time (under an hour), compare the plants and choose a winner. If time allows, declare this last event an information-testing segment and repeat the contest to give each child a chance to produce a winner. This also gives you a chance to see who might be missing the point. Another variation is to have a contest to prevent wilting. Look at pictures of desert plants and point out adaptations to dry habitats such as small leaves, light and reflecting surfaces, waxiness, fuzziness, or oils to slow evaporation. How might the adaptations be translated to the celery stalks?

76

Using Leaves in Art Projects

In the following projects, leaf identification is not the goal. The children will be handling various leaves for their aesthetic effects and, in the process, will be exposed to many distinctions that will make learning identities a simpler task. As you work with the children, use the proper common names in association with the textures or colors the kids have noticed.

PRINTS

The leaves can be painted with poster paint and placed paint-side down to transfer their shapes onto paper. Experiment with the differences in quality of the upper and lower surface of the leaf, and with using the same leaf to create repetitive designs. To display differences in color intensity, use the same leaf without repainting. You can do the same things with ink printing. You can use a standard stamp pad, or make your own with moleskin and food coloring. Glueing the leaves onto stiff paper or Styrofoam may help keep the surface level for even inking. For both printing processes, allow for many scrap-paper experiments and plenty of "good" paper for a number of "proofs."

SILHOUETTES OF LEAVES FOR STENCIL OR OUTLINING

Trace the outline from an actual leaf or shadows of leaves on a paper. The cut-out leaf can be used as a design element on another piece of paper, and the opening it left can be used as a stencil for making multiple copies of out-lines or colored-in shapes. Color using crayons, sponged-on paint, or spray paint. The stencil can also be used to make splatter-prints, by splattering poster paint evenly over the stencil where it is placed over the paper (or fabric) to be decorated. A satisfying spray can be achieved by holding a paint-dipped

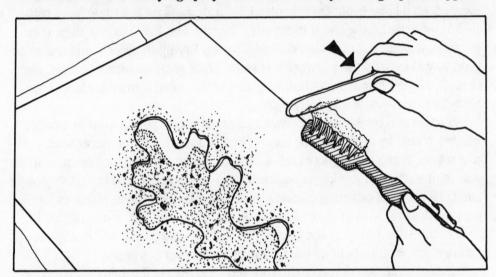

old toothbrush, bristles up, over the stencil and rubbing it briskly with a pencil-size stick or ice-cream stick. The amount of paint, its thickness, and the types of strokes give different effects, so provide material for experimenting. Other elements (sticks, leaves, or flowers), when laid directly on the paper and splattered to get their shadow effects, will give pleasing silhouettes.

COLOR SMEARS

Leaves and flowers will produce a colored smear when rubbed on card-weight paper. Obviously, you need to do some soul-searching before presenting this activity, as you will completely destroy the plant parts and they must be fresh. Using end-of-summer material or garden plants is the easiest to justify, but give some thought to the destructive aspects of this exercise. At least make the children aware of your concern and invite them to help with the decision.

■ Use the colors of the leaves and flowers to re-create a resemblance of the plant. The smudges are hard to control, so the kids may prefer to draw the outline first in pen or pencil.

■ Make an "impressionist" portrait of a landscape, using plant material that is predominant in the view.

■ Collect predominant colors of various environments with smear lists. Many of the colors you collect by smearing will fade in sunlight. Test the colors for fastness by shading part of your sample or by making two sets and exposing one to sunlight. Several sets would let you test the relative speed of fading over longer time.

PRESERVING THE LEAVES THEMSELVES

Colored leaves are more eye-catching, but the greens are interesting, too. You can preserve leaves fresh out of the collecting bag or after flattening and pressing them between sheets of newsprint or construction paper weighted evenly with a large book. Once pressed, their dryness makes them more brittle but more satisfying as art materials. They are less likely to lose their color if mounted when dry and will print inks better. Drying the specimens is also a good way to expand the activity into two different sessions, one to collect, one to craft. However, young children may lose enthusiasm if you wait more than a day before using their collections.

Leaves can be preserved between layers of waxed paper, even in a paper sandwich bag, by arranging the leaves then sealing the waxed paper with a warm iron. Using two layers of self-adhering plastic is even more permanent and impressive, but it takes experimentation and practice to do effectively. A single layer of self-adhering plastic preserving an arrangement of leaves (or pressed flowers) against colored paper is also attractive. Use a dab of glue to hold each leaf in place, as the plastic has a static charge and will totally disarrange any arrangement as the plastic is placed over the paper.

Select materials that give the children the greatest flexibility and opportunity to do impressive work on their own. For instance, small pieces of plastic are easier to handle than large ones. You might have a variety of sizes precut for their needs.

Self-adhering plastic does a nice job of mounting specimens on all kinds of paper: cards, poster board, colored papers. The samples can then be used for:

■ Flashcards in identification games.

■ Transformed card games such as Concentration. (Collect pairs of small-size leaves, press and mount them on same size of cards. Turn them facedown. The children take turns flipping any two at a time, searching for matching

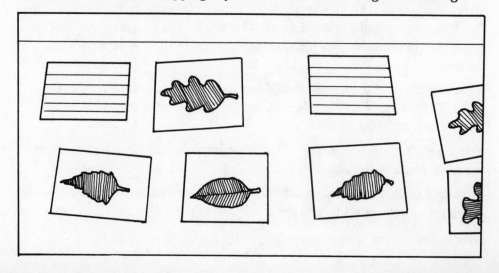

pairs. Prereaders are aces at this game.) Also consider transformations of Old Maid, Slap Jack, Go Fish, and so forth.

- Sets of leaves to describe various habitats, or seasonal changes.
- Growth stages of a particular plant.
- Greeting cards.
- Placemats, name cards, bookmarks.
- Decorative coverings on flat or curved surfaces (wastebaskets, pencil holders, notebooks . . .).

77

Response to Stem Damage

We become so used to the plant habit of growing upward that we seldom stop to wonder how a plant knows which way up is. The determination to grow up and out is so strong that when the stem of most plants is damaged, side shoots continue the outward growth. This consistent growth away from the central stalk or trunk is the result of the cellular stimulation of a plant hormone called an auxin. Auxins that are produced in the stem of a plant accumulate in the tip ends of the upper plant. In a tree, the larger, terminal buds have the greatest concentration of auxin, to the point where the auxiliary buds are often inhibited from growing. If a terminal bud is damaged, auxiliary buds will accumulate the auxin and begin to grow.

Choose several specimens of actively growing plants of approximately the same age, size, and species. Mark and leave alone a plant or two for comparison with the plants you will be treating. Then cut off segments of the tallest of the growing stalks on some plants and the side stalks of others. Label the plants as to the kind of damage, current height, and the date. How do the side shoots respond when the main stalk is damaged? How does the main stalk respond after the side shoots are removed? Compare the spaces between the leaf stems to measure relative rates of growth.

Collect pictures or drawings of trees in your area that show responses to past damage to the main branches or terminal buds. Conifers will clearly show crooked trunks where side shoots once took over after removal of the dominant stem. Many multiple hardwood trunk trees of the same size indicates that lumbering occurred. Investigate areas where brush has been cut away within the last year. When a tree or shrub is cut, the auxins will stimulate buds to form around the rim of the stump. A dense growth of suckers will grow from the buds. Eventually most of the suckers will lose out in the competition for nutrients, and two or three will stay as the main trunks.

78

Does a Plant Need to Breathe?

By keeping the soil saturated with water around the roots of a potted land plant, you will quickly answer the question of whether a plant needs to breathe. Any living tissue must be able to exchange waste gases for fresh elements in order to live and grow. A plant with green leaves has the impressive ability to provide itself with the food it needs. Sealed in a glass container, a houseplant can maintain itself for a long time. During the daylight hours, the

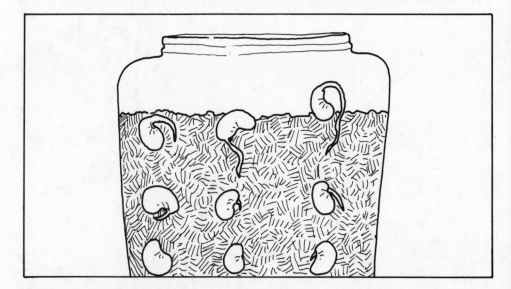

leaves absorb carbon dioxide from the air, break it up and reassemble it as sugars in photosynthesis, and then dispose of the waste, oxygen, back out of the leaf. During the dark hours, the oxygen is reabsorbed to help in the growth process, where the sugars are "burned" to power the cells to grow. In this process, the waste gas is carbon dioxide. Drowned roots in soggy soil would not be able to function and the plant would die.

For sprouting seeds, oxygen is essential for the great growth and expansion that occurs while the plant is too young to produce its own leaves. This can be demonstrated by filling two equal-size jars with large presoaked beans, such as lima beans or peas. The large size will provide some air spaces so that the growth can begin easily. Seal one of the jars and cover the other with a moist wad of paper towel or a cloth to keep the seeds from drying out. Make sure that the children understand that air can get into one set of seeds but not to the other. Observe the changes over several days. Do the seeds in the sealed jar grow at all?

A similar experiment can be done by planting seeds at various depths against a glass side of a container filled with potting soil.

The deeply planted seeds will have access to less oxygen than will the seeds that are planted near the surface and should show some difference in growth rate. Encourage speculation as to the effects of deeper planting as the experiment is set up. Keep a list of possibilities as the seeds grow and list causes when the effects are noticed. When the deeply planted seeds reach the surface, is a difference in growth rate still apparent?

79

The Holding Power of Roots

If a plant is subject to grazing animals, it is to its advantage to have a root system that can resist the tug of its leaves being pulled off, as the roots can then grow another set of leaves. Many of our lawn and field plants have evolved this tactic, and you can test for "root power" on any lawn. Especially of interest are those lawns that support a variety of plant species. *It is essential that you check with the owner* before you let a group of kids loose on a neighbor's lawn.

Bring along a trowel to dig out those champion roots that give up their leaves and stay behind. Tug away at various plants for a short time, then decide on a grading system for root resistance. The system might be simply a three-part range of: easily pulled out, harder to pull out, and have to be dug out.

Once the children are looking appreciatively at the roots, challenge them with these questions:

■ Do any of the roots show areas of food storage for next spring's growth? Dandelions, wild carrots (otherwise known as Queen Anne's-lace) or wild onions all do.

■ Do any roots reveal techniques of colonizing the territory with underground stems that come up nearby as new plants?

If you happen to pull up clover plants, check for the little nodules created by the growth of nitrogen-fixing bacteria. These bacteria feed off the plants as they help the plants to grow by processing nitrogen from the air into a form usable to the clover and other plants. You may notice that grass is often greener by the clover patch.

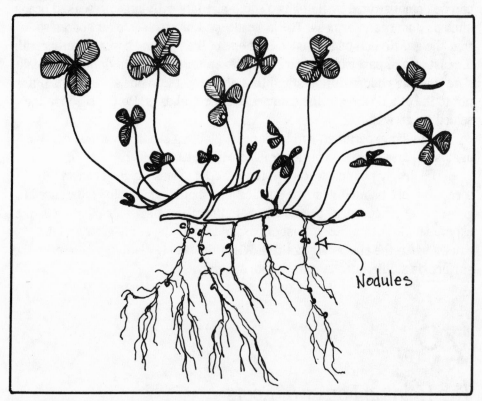

Nodules

Provide materials for colorful displays of specimens (or drawings of the specimens) in their various categories. Once the superholders are well known, the kids might want to stage a Plant Olympics with prime contestants of their choosing. Encourage personal names and banners to dramatize the contest.

80

Carrot Machine

If cucumbers are covered with a saltwater solution, they lose cellular water and become limp by a natural process called osmosis. Limp carrot sticks can be made stiff again by placing them in fresh water. The cells become filled with fluid, again by osmosis. How can osmosis be responsible for both events? Osmosis is the movement of water across a cell membrane; movement occurs from a greater concentration of water to a lesser concentration. In the case of the cucumbers, the salted water is "diluted" with salt so it is a lesser concentration. Water within the cells of the cucumber must move out into the salted water. For the carrots, the purer concentration of fresh water must move into the slight salt solution within the carrot cells. Osmosis is a natural process and an essential one. Water moving from the soil into the roots by osmosis maintains all plant life on land. Nutrients are carried along into the plant as the water moves into the roots.

You can build a device to measure the pressure of osmosis by building an osmometer, using a carrot and a straw. When a carrot is growing in the ground, its cells are usually absorbing water by osmosis and then passing the water on up to the stalks and leaves aboveground. If a clear straw is substituted for the leafy tops, the fluids from within the carrot can be observed rising under osmotic pressure. (See illustration for instructions.)

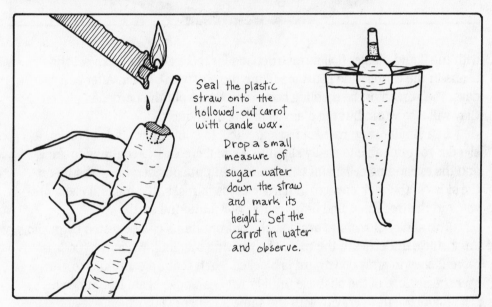

Seal the plastic straw onto the hollowed-out carrot with candle wax.

Drop a small measure of sugar water down the straw and mark its height. Set the carrot in water and observe.

For Discussion

■ In the celery-tinting experiment (page 83), the leaves functioned to pull water through the plant. As there are no leaves here, how does the water get into the carrot system? (Osmotic pressure is a kind of a "pump" for plants. Most of the work is done by the leaves in a growing plant. You will notice that the carrot pump can only go so far.)

■ Try tinting the water with dark ink or food coloring and then halving the carrot at the end of the experiment. Can the effect of the ink be seen throughout the carrot, or are there main veins by which the water flows? What might be the function of the other cells in the carrot? Are they colored at all?

81

Responses in Roots

Root strength can be impressively displayed by planting several beans (which have been presoaked overnight) in hardening plaster of Paris. To make sure that the beans are in the center of the plaster mass, fill a paper cup halfway

with the liquid plaster (follow instructions for mixing on the box), set the beans in the center, and continue filling the cup with plaster. After several days, the growth of the seedling beans will have split the hardened plaster. Clay will also work, but the effect is not as dramatic.

In a similar way, roots of trees are able to split softer kinds of rocks. A slender root tip, able to easily slip into a small crack, will continue to grow, and the expanding cells, stiff with water, can put considerable pressure on the sides of the rock crevice. In harder kinds of rocks, the mineral elements win and the tree must find other routes down into the earth.

The ability of roots to avoid such confrontations can be tested by placing an obstacle underneath the roots of a growing seedling. When the root's natural tendency to grow downward is blocked, each root tip continues to grow over the surface of the obstacle until gravity can once again lead it down.

Roots orient to gravity with the same sureness with which growing leaves head into the sunlight. Start some easily grown seeds such as radishes between sheets of moist paper towel. Once the roots begin to grow, mount the seeds against a clear piece of plastic, such as a cover for take-out salads, keeping the towel behind the seeds moist and firmly pressing the seeds against the plastic. A piece of cardboard cut to fit tightly will keep the parts

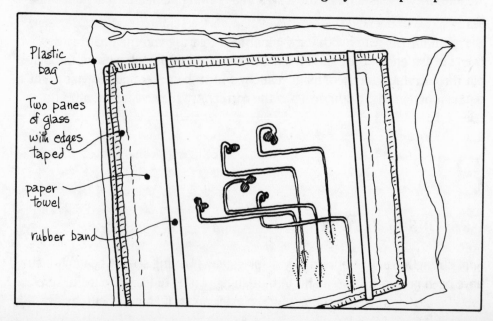

together. Stand the mount on its side until the roots have stretched out and oriented downward. Then rotate the mount ninety degrees to one side and let the roots grow some more. The roots should turn down ninety degrees. What do the children predict for the leaf stem? This unerring inclination to grow in the appropriate direction is accomplished by auxins. Root and stem auxins seem very susceptible to the pull of gravity and are found in greatest concentrations on the lower side of a horizontal stem. If the stem is a leaf stalk, the auxin stimulates the lower cells to elongate, which causes the stem to curve upward. If the stem is a root, the lower cells are inhibited by the auxin, which results in a curving down of the root as the upper cells grow at a normal rate.

Challenge the kids to figure out a way of calculating the total length of a plant's root system. Use a well-grown potted plant to be sure of counting all the roots. Rinse off all the soil so the roots are exposed and encourage the children to figure out shortcuts. One method that works if all the roots are about the same general width: measure, then weigh a length of root; find out the total weight of all the roots and divide by the weight of your known length; that number times the length of the known piece will give an approximate length of all the roots.

Find out which parts of the plant weigh the most. Guess before measuring whether the aboveground parts weigh more or if the roots have the greatest mass. Does the relationship vary during the plant's various life stages? Is there a difference between annual weeds or flowers and perennial plants?

82

Some Food Crops That Also Serve as Storage Areas for Plants

Many of the swollen roots or stems of plants are useful to us as food because the plant has stored its food supply in these areas. Often the food is stored over the winter. In the spring, the warm earth will initiate new growth. Many plants flower and set seed during this second season of growth. Any supermarket with fresh produce will have a good supply of these kinds of vegetables. Carrots, parsnips, turnips, onions, garlic, potatoes, yams, ginger root, and radishes are common examples.

If you have space outside, you can plant these vegetables when the soil is warm (once deciduous trees are growing leaves) and then observe second-season growth. In a classroom or a sunny window, the plants can be watched more closely by placing the cut-off top inch or so (from the stem end) in a shallow bowl filled with water. The water and warmth will stimulate leaf growth and the stored foods will nourish the new leaves. When the food runs out, the growth will stop, as there is no soil to supply more nutrients for growth. (Would twice as much root produce twice the growth?)

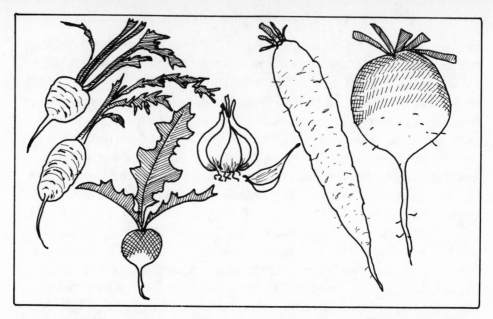

Most of the plants we grow for their nutritious leaves are annuals. They will grow, flower, and make seeds in one summer. The plants that give us stored vegetable foods to eat are usually biennials. They flower the second year, set seeds, and die. Potatoes and ginger tend to keep growing up from the same area year after year if unharvested. Both these plants are stems that are used for underground storage. Sweet potatoes, white potatoes, and ginger will grow handsome plants if potted in standard growing soil, just under the surface.

83

A Rose Is a Rose Is a Cabbage Bud

A budding rose, an autumn cabbage, and an artichoke are all buds, in that they all consist of modified leaves, growing in a compressed spiral around a central stalk. In the leaf bud, the stalk would eventually stretch up, raising the leaves to the sunlight. The cabbage is in fact a leaf bud, but the stalk is permanently stunted and the leaves are dense with stored nutrients. A bunch of celery is a similar horticultural anomaly, but only the stems store the extra food. A flower, such as the artichoke, is also a compressed and modified stem of leaves. It is a leaf bud turned into a flower bud by the plant's own chemicals.

These similarities in form and function are complicated to explain, but they are obvious when explored firsthand. In the early spring, when tree buds are just beginning to open and show their leafy interiors, collect a red cabbage, a bunch of celery, head lettuce, an artichoke, and any tree or flower buds that seem appropriate. Have enough extras so that they can be opened in various ways: from leaf-by-leaf picking to slicing, both crosswise and along the stem.

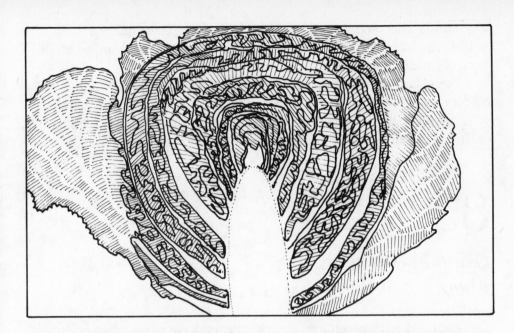

As the group investigates each plant, keep track of the words used to describe the arrangements and relationships within each "bud." Do the relationships and words used for one bud transfer to another, or help to see the parts in the other? Which words can be used to describe patterns in all the plant parts?

84

Paperwhites: Growing Spring Flowers from Bulbs Indoors

Another budlike form is the bulb of a paperwhite narcissus, which is also available in yellow forms. It is a great satisfaction to grow these pleasant flowers from a dry-looking bulb, and the process is very simple. Modern forms of the bulbs, available at garden stores or through garden catalogs, can be started simply by settling the bulbs, pointed end up, in a bed of stone chips or pebbles. Keep the water level up to the base of the bulb to encourage root growth. The dense, white roots usually appear first and take a firm hold of the stones. The slender, green leaves appear soon after. The flowers will be blooming about a month after starting, depending on the warmth of the room and the constancy of the watering. They can be grown in soil in pots, but even if they are carefully dried out after the leaves die back and are kept from freezing temperatures, the next season's bloom are not as full the second time around. (This is an excellent activity to start and then give as gifts for others to enjoy.)

Try starting a number of bulbs at the same time and then growing them under different conditions to compare effects. Different temperatures, amounts of light, and humidity can often be found within one room. Leaves and flower stalks can be preserved easily by pressing them under a book weight. They can then be mounted as a chart to express their various conditions of growth.

Onion bulbs are very similar in structure to narcissus bulbs and the milder ones can be dissected to observe the thick stems and spiral pattern. What would be the advantage to having strong oils in a bulb? Try growing onions in the same way as the narcissus. You could get some flowers.

85

Using Information About Plant Parts in a Game

Have the children draw outlines of each other on large pieces of paper, either using silhouettes cast by a strong light or by directly tracing the outline of a person lying on the paper. Suggest that children assume "plantlike" shapes for their outline, then let them modify the outlines somewhat as they draw to increase the resemblance to plant structures. The shapes can then be colored in appropriate hues to differentiate the structures of leaves, roots, stems, flowers, and fruiting parts. Encourage discussion on what kinds of habitats these new plants might be best adapted to inhabit.

86

Bring Trees into the Classroom

TREE #1

Trees are big, and it is fun to capitalize on this attribute in the creation of group projects. The cheapest sources of plain paper for a background are the rolls of newsprint from newspaper printing companies or the folded sections of unused billboard copy, often free from a company that handles billboard advertisements. Powdered poster paint is also a way to save money. Draw the outline of the group's favorite tree (or do several mixed together) with chalk and paint away. Try to be as realistic as possible in coloring for various parts for an authentic look and for a greater awareness of what the tree actually looks like. The large body movements of slapping the paint over a large area is the most fun. Chunks of fine-grained sponges, cut into sections the size of bread slices, will do nicely for paint brushes.

TREE #2

Make a large tree made of tree parts. Go out on a scavenger hunt for tree parts: leaves, twigs, branches, bark, wood, and (unattached) roots. Discuss tree parts before going out and make it clear that no living part may be collected. Provide large bags to ensure lots of supplies.

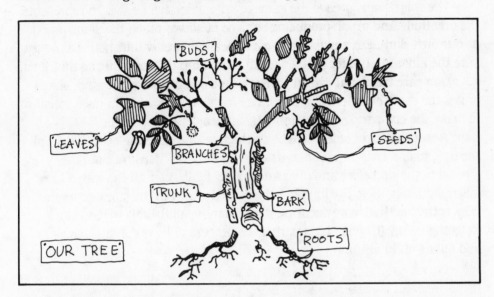

Assemble and sort through the material, then put the parts together in a tree shape. This can be easily done on the ground. As part of a more permanent display, the pieces can be glued to sections of cardboard (from an appliance store, perhaps). Do the initial glueing on the ground. It is also possible to tack material to a bulletin board. Have some long tacks to nail up the thicker pieces.

The tree can then be the center of seasonal art exhibits (making appropriate changes in the tree). It can also be the basis of discussions or discoveries on the functions of the various parts, their role in tree growth, and their connections to the various other living things in their environment. Let the children express these ideas in drawings or models, which can be added to the tree exhibit.

87

How Many Leaves on a Tree?

Count the number of leaves on some easily identified unit of a tree, such as a small branch with a cluster of leaves on the end. Estimate the number of branches of that size that make up the tree and multiply by the number of leaves on the initial branch.

This figure could be made more relevant to understanding tree biology by using the number of leaves to compute the surface area of the tree. Choose an average-size leaf, then figure out its approximate area by tracing the outline onto graph paper and counting the squares that are covered at least half by the leaf. Multiply this number by the number of leaves. This gives us the total "solar panel" surface of the tree. Compare the solar surface relative to number of leaves in a tree in full sun and a tree in complete shade. If the number of leaves is similar, is there a difference in sun-gathering area?

The solar-panel area is the same as the undersurface, which represents the breathing and vapor-producing surface (it serves much the same function as our own skin). Put a plastic bag over a leaf for twenty-four hours and measure the amount of water (or its weight) given off by the tree during that time.

Compare trees that live on sunny hilltops with trees of lowland, shady areas, the same or different species. Use trees of similar height (as a constant) to make the comparison more dramatic and easier to appreciate.

To see the effect of leaf weight on the tree, measure the exact height of the tip end of a low, thin branch using a plumb line. Measure from the ground to the tip before and after autumn leaf fall. After all the leaves have fallen, add weights to the limb to bring it back down to the summer level. Why is this method more accurate than catching all the fallen leaves and weighing them? (Drying out is part of the process of leaves changing colors and falling off in autumn.)

88

The Autumn Leaves

If you happen to be with children on a day when leaves are falling, spend awhile simply running around catching the leaves before they hit the ground. The kids may want to make up fancy ways of catching them once they get good at standard efforts. The leaves can be used in a great variety of ways. Besides the following games, the kids will be able to make up others.

■ Sort the leaves according to size and to kind. Find leaves that are the farthest from their parent tree.

■ Use the smallest leaf as a unit for measuring the other leaves.

■ At what times or during what circumstances do the leaves fall the most?

■ Rake an area. Then wait ten minutes and count how many leaves fell within that time. Multiply to figure how many will fall in an hour. Count all the leaves in another area of the same size and figure how long it took them to fall.

■ Try to figure out how long a leaf has been on the ground by its dryness.

■ How many leaves does it take to make a stack an inch high? Do different kinds of leaves take different numbers? Figure out the thickness of one leaf by the number it takes to make an inch.

■ In a carefully controlled situation, weigh a leaf, burn it, and weigh the ashes. Is there a great difference between freshly fallen leaves and ones that have been on the ground several days?

■ Make a great, thick pile of leaves and practice jumping in (having made sure that only leaves are in the pile, no sticks or other people).

89

Estimating the Height of a Tree

METHOD #1

In the method, you will use three known measures to find the unknown height of a tree.

The day must be sunny enough to cast shadows, and the shadow of the tree must lie along flat ground. Measure the shadow first. To make the figuring simpler, use only inches or centimeters. Next, dangle a yardstick (or meterstick) straight down until it just touches the ground. Mark the distance covered by the stick's shadow and measure it. The formula for finding the tree's height works this way: the relationship between the tree and its shadow is equal mathematically to the relationship between the stick and its shadow. Fill in the numbers (using the same unit of measure) of this formula:

$$\frac{\text{tree's height}}{\text{tree's shadow length}} = \frac{\text{measure stick (36 inches or 100 cm)}}{\text{stick shadow length}}$$

From this basic set-up, you do some algebraic moves and come up with:

$$\text{tree's height} = \frac{\text{stick length x tree shadow}}{\text{stick's shadow length}}$$

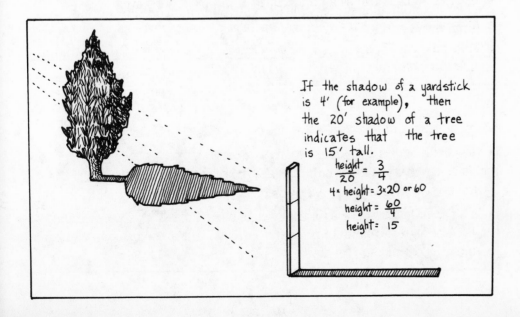

If the shadow of a yardstick is 4' (for example), then the 20' shadow of a tree indicates that the tree is 15' tall.

$$\frac{\text{height}}{20} = \frac{3}{4}$$

$$4 \times \text{height} = 3 \times 20 \text{ or } 60$$

$$\text{height} = \frac{60}{4}$$

$$\text{height} = 15$$

Or: The stick's length, divided by the length of its shadow, and then multiplied by the length of the tree's shadow, equals the tree's height.

Suggestion: Try out your figuring first on a small tree that you can measure with the yardstick. That way you can test the formula and proceed with more confidence.

METHOD #2

This method works well with younger children. Stand far enough away from the tree so that you can see its entire length without moving your head. Hold a straight stick or pencil at arm's length so that the top of the stick appears to touch the top of the tree. Put your thumb on the stick where it matches the foot of the tree. Without moving the hand any closer to your eye, turn the stick so that it covers the ground horizontally, your thumb still touching the stick and appearing to touch the base of the tree. Have a friend move to stand in the spot covered by the tip of your stick. Then go and measure the distance between the friend and the foot of the tree. It will equal the distance between the top and the foot of the tree.

90

Looking into Tree Buds

It sometimes surprises people that buds can be found on trees by the end of summer. The tree makes the buds during the warm months when the most sunshine is available for growth. The buds are dormant during the winter (or dry months), and ready to open when stimulated by the warmth, water, or the increasing day length of spring. Many species of northern trees and shrubs must have cold for a certain period of time in order to "set" the buds. This need for a conditioning process will keep the buds from opening during autumn warm spells.

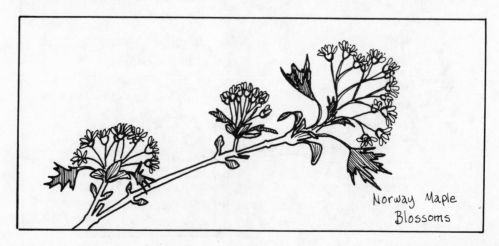

Norway Maple
Blossoms

At any time after December, tree twigs can be brought indoors and "forced," that is, put in water as you would flowers and allowed to open. Some buds may open to reveal flowers along with leaves. Apple, dogwood, or cherry branches will bloom beautifully. Other tree flowers are less conspicuous but are equally fascinating. Maple, oak, birch, and willow branches will bear flowers that are pollinated by the wind. Not needing to attract insects, their colors and forms are more subtle. Don't overlook the miniature intricacy of the unfolding leaves.

Buds can be forced another way. In spring, as the leaves inside the buds begin to push apart the protective scales, the buds can be taken apart easily. Small fingers do a good job; children may enjoy using sturdy pins or toothpicks to pick the bud apart more carefully. Notice the arrangement, sizes, and numbers of the leaves. Taking apart buds of a variety of species or a number of the same species over a period of time enhances an understanding of the growth patterns of leaves on the summer tree.

91

Twigs

The twigwood on the ends of the branches on most trees holds clues to the growth rate of that branch. Look along the smooth bark of a deciduous tree or bush, starting at the tip and moving back toward the trunk. At some point, from a fraction of an inch to a foot or more, the bark color will change noticeably. Where it changes, look for several fine lines, close together, encircling

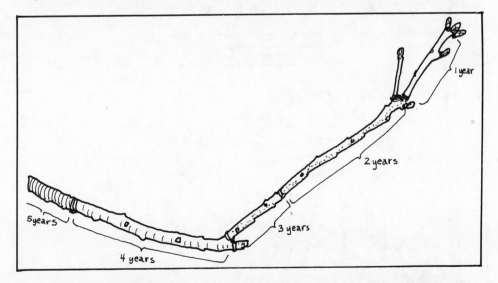

the twig like a series of bracelets. These lines mark the position of the terminal bud and the end of the previous year's growth. Within that terminal bud were the basic structural elements for the now-extended twig, leaves, and new bud, all of which now stretch from the bud scars to the tip of the twig. Keep looking back along the twig until you find another color change and bracelet scar. This scar marks the portion of twig that grew the previous year. Varying lengths of annual growth indicate the effects of various growing conditions. Setbacks may be caused by pests, cool weather, or disease. A growth spurt might indicate an increase in light (the loss of an over-shading tree) or a summer of rain.

The structure of the most recent twigs is usually very symmetrical. The basic pattern of the tree species is clearly expressed. With every year of growth, the pattern is increasingly altered by environmental influences. As the twig becomes branch, the bark becomes thicker, until the evidence of the yearly growth is obscured.

Compare the annual growth on twigs on the same tree (the branches in the sun should grow the most) and on twigs from different trees. Do they show similar rhythms of growth over several years?

92

Finding the Ages of Young Pine Trees

A pine tree that bears its whorls of branches all the way down to the base will reveal its age in the number of whorls. Examine the tip of any branch, and you will see the pattern of one leader branch surrounded by a whorl of secondary branches at its base. The leader and the whorl all grew during one year from a cluster of buds. Therefore, each leader and whorl together equals one year, whether grown on a side branch or on the main trunk.

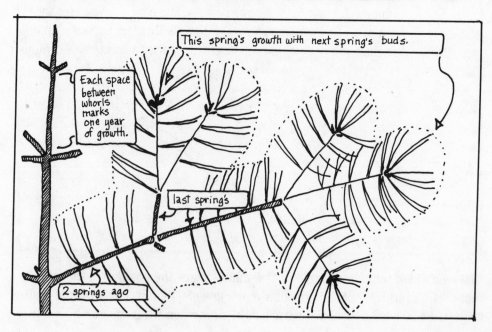

Starting at the base of the tree, begin counting at year four, the approximate age of a young pine when it puts out its first whorl of side branches. Then add one for each junction of leader and side branch (sometimes there is only one, or a stub of one, left). If the tree is too high to continue counting easily, keep the count going out along a side branch. To the tip of the branch and to the top of the tree is the same count.

If there are child-size pines around, encourage the children to find pines their age or perhaps to find the whorls that represent the year of their birth. You may need to help them understand why they must count down from the top where the more recent growth has occurred.

While looking at that special year, you can tell something about how good a year that was for the tree by the relative lengths of the leader spans. Extra inches between whorls indicate increased access to sunshine, moisture, or nutrients. A shorter growth might result from insects eating the leaves, drought, or extended cold (but never "a short year," as one logical student put it!).

93

Bark Cork

Tree bark is produced by a special layer of "mother cells" that cover the wood of the tree, just under the old bark. In the spring, new bark cells are made that fit over the new wood cells, protecting them from invading insects or fungus and keeping the new wood from drying out. The bark of a Mediterranean cork oak excels in this ability to block the passage of moisture and is grown in groves for commercial use. At regular intervals, the cork is harvested by stripping the oak of its corky outer bark. The bark is then perforated with a punch. The punched-out segments are the corks, now effective seals for fluid-filled bottles.

Cork is an interesting material to investigate. Save some wine corks or buy some corks at a wine-maker's supply store. The spongy cell walls of cork allow compression but spring back when released. Cork is somewhat difficult to cut with any but the sharpest razor or knife, so cutting activities are not recommended for small children.

In most corks, especially wine-bottle ones, a series of parallel lines or layers can be seen running longways through the cork. These layers represent the annual growth rings, similar to those found in the wood. If you do any cutting of the cork, you will find that the ease and cleanness of the cut depends on the angle of the knife relative to the rings. Have other samples of bark cross-sections to compare with the cork. All bark cells have cork inside their walls, to varying degrees. Trees that have adapted to live in very dry conditions, especially oaks, tend to have thicker layers of cork in their bark.

94

Hardwood/Soft Wood

A soft wood such as pine is useful for house-building because of its relative lightness and strength. A hardwood, in comparison, is used more for flooring or furniture. Oak wood is a native hardwood unsurpassed in weight and durability. In general, all conifers are referred to as softwoods. They tend to grow more quickly than deciduous trees, and their green leaves can make food whenever conditions allow. All trees that lose their leaves in winter are termed hardwoods, even very soft-wooded trees such as aspen or catalpa. A hard-wooded tree is either a very slow growing plant, like a hornbean tree, or has special cross-grained wood fibers, as do oaks. Pieces of wood can be tested for their relative hardness, or reduced to equal size and weighed. The different sizes and densities of woods relate also to resonance and tone. Slabs of wood will sound when struck with a wooden mallet. The effect is enhanced by suspending the wood on a string or supporting the pieces on sections of dense foam strips. With careful experimentation, a full scale of tones can be achieved, but rhythmic fun can be had with only a few distinct tones. Try to find a xylophone for the children to play with. "Xylo" means "wood," as does "xylem," the cells that produce the wood of a tree.

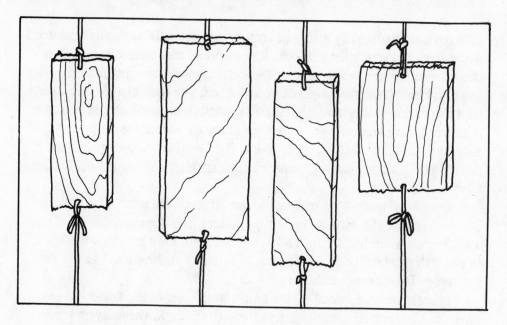

95

Explorations Using Lumber and Cross-Sections of Wood

If a tree has been cut down in your neighborhood, try to get a full cross-section of the trunk. Better yet would be to take your group to the site so that they can get a sense of the tree's size and the process of taking it down. Consider using the stump, slices of the log, and sections of the branches in the following ways:

■ Count the age of the tree. Light-color rings indicate the more rapid growth of spring and summer; dark-color rings mark the slower, denser cell growth of late summer and fall. With very old trees, the annual rings tend to be close together and hard to differentiate, especially in the outer rims. Count back from the edge and find the years in which the children were born. Figure out other historical dates such as the building of the main town buildings, the founding of the town, and so forth.

■ Encircle the circumference of the trunk with a string and mark and measure its length. The string can then be used to find other trees of approximately the same age, especially if they are the same species in a similar habitat.

■ Obtain samples of different kinds of woods from a lumberyard. A large yard may have a bin where workers throw extra pieces they no longer want. Ask if you may have access to the bin, and try to get the pieces identified. These samples can be used to test hardness (by pounding nails into them) or heaviness (by cutting them to equal size and floating them or weighing them). Have a bar of floating soap and chunks of metal handy to experience their extremes.

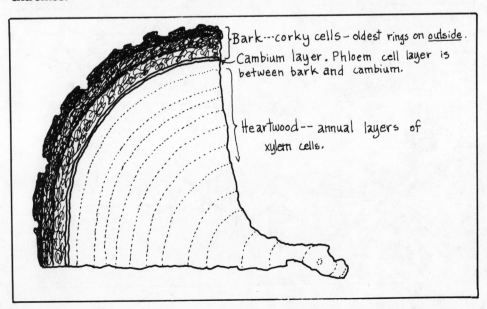

Bark---corky cells - oldest rings on <u>outside</u>.
Cambium layer. Phloem cell layer is between bark and cambium.

Heartwood-- annual layers of xylem cells.

■ Find out what the children already know about uses of woods. They may have heard people talk about furniture made of maple, pine, oak, or cherry. If possible, show examples of wood used for specific purposes. Ash is used for baseball bats; persimmon for golf clubs; oak for flooring, stairs, and mallets; and maple and birch for cutting boards and salad bowls.

96

Comparing Leaves of Deciduous Trees and Evergreens

The textures of leaves tell the story. You might want to introduce the two kinds of leaves by feel at first. Place examples of both in bags or boxes with hand-size holes for easy access. Ask the children to decide which leaves are tougher, larger, moister, or slicker. All these attributes are important to the relative adaptations of both kinds of leaves. Keep a list of the terms the kids use to make later distinctions between evergreen and deciduous.

The evergreen leaves have tough leaf skins or cuticles (and often reduced surfaces). These prevent excessive evaporation of water during the months that water is frozen in the ground. The upper sides are often slick with a waxy coating. This helps the trees shed snow and prevents the branches from being weighted down and broken. Pine needles are usually slick in one direction and rough in the other. The underside of both kinds of leaves is often much lighter than the upper side. For the evergreens, especially, lighter green coloring will reflect sunlight, which cuts down on water loss from excessive radiant heating. On trees from climates where snow might reflect light onto the underside of leaves, that surface can be almost white.

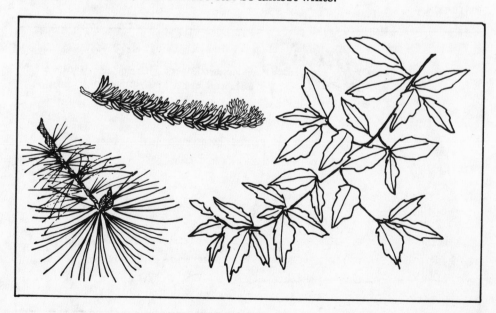

In spring the deciduous leaves are dependent on water availability to expand their leaves. As the more tender type of foliage, they are also more susceptible to insect predations. This also means that a deciduous forest supports a greater array of animal life.

Discuss these distinctions with the children and ask them what experiments they might set up to test the difference between evergreen and deciduous samples.

Some possible experiments are:

■ Leave cut branches of approximately the same weight out on a table for an hour or so. Which looks the most wilted? Is there a weight loss overnight?

■ Put two equal-size cuttings in containers with measured amounts of water. Check the water levels after a day to see which specimen uses the most water.

■ Leave sample specimens of both kinds of leaves in an area where they will have a chance to rot. Mark the area so the leaves can be identified later. After a month or so, check on the results.

97

General Instructions for Growing Plants in Pots

WHAT KIND OF POT?

If the plants will have to take care of themselves for the most part, do not use clay pots. Plants in clay pots will need more watering, more fussing over. Clay pots are good choices if you are (1) working with cacti or succulents or (2) planning to sink the pots in a garden bed. A nonporous pot will let you water less frequently, is lighter, and is easier to handle. Many kinds of containers can be converted from waterproof liquid holders to plant pots by punching holes in their bottoms. Do not allow the prospect of infrequent watering lead you to skimp on the number of holes you make. Drainage is very important. Roots in soggy soil will literally drown. A hole every half inch is about right. Put holes in the lower sides of a deep container. Some potential containers include: all sizes of milk cartons, loaf pans, pressed-foam coffee cups, metal cans, and margarine tubs. When you are starting many seeds, even egg cartons and little paper cups are adequate, but plan on handing out the seedlings for home or transplanting them elsewhere after about two weeks of growth.

WHAT SIZE POT?

As with the type of pot, the choice of size involves considerations about the problems caused by too much soggy soil in a large pot and too dry a soil in too small a pot. The right proportions of soil to seed or seedling is something you get a feel for after you have handled a number of plants (and made a number of mistakes). If you and your children are just starting out in the practice of horticulture, you may wish to begin with an experiment

using the same seeds and the same soil in different sizes of pots to see which works best. If you use seeds from seed packets, there will be instructions on the back of the packet to guide you. If you are transplanting small nursery-grown plants from flats or six-pack containers, move them to a pot that will let you put a half inch of new soil in the bottom and a quarter inch around the sides of the plant's root ball or root length. Start seeds in pots or flats that hold two to three inches of soil. Any less soil will dry out quickly, killing the vulnerable seedlings.

STARTING SEEDS

How deep to plant? Again, this can be one of your experiments. In general, the gardener's rule is to cover the seed four times its width with soil. For long or flat seeds, use the measure of its narrowest width. Some plant seeds, such as those of the lettuce family, need sunlight to stimulate sprouting. Check the packet or a book on gardening. Seeds of other plants, especially those of dry climates, such as black locust and lupines, have tough seed coats to protect them from drying out, and they need to be filed or nicked to speed up their growth response to moist soil. If you are interested in trying to grow native plants from seeds, you may have to expose seeds to periods of near-freezing cold before they can sprout. A good book on native plant horticulture is listed in the bibliography.

Once you get the seeds planted, they need to be moistened to start their growing processes. A blast of water from a hose or watering can is likely to churn up the soil and float the seeds to the surface. A hand-held misting bottle produces a sufficiently gentle spray, but is very tedious if there are many pots. Placing the pots in a tray of water so that the water level and the soil level are the same will soak the soil evenly in an hour or so. Lift the pots out when the surface looks wet and let them drain thoroughly.

As the surface of the soil will be the first to dry out, you can keep the seeds moist by covering the pot with a piece of newspaper cut to fit the soil surface exactly. A plate of glass or a plastic bag tent will also retain the moisture. Uncover or raise the cover as soon as sprouting begins. Moist, still air around the tender seedlings will promote "damping off," the growth of fungus and the death of plants. Sunlight isn't necessary for sprouting (except for lettuce), but warmth, especially from the bottom, is helpful. Be careful not to cook them; keep the temperature around 70º F.

WHAT KIND OF POTTING SOIL?

The easiest source of potting soil is a garden center or plant department in a home supply store. A general-use potting soil can be bought already mixed and sterilized, so that the risk of damping off is minimized and there are no weed seeds or pests to damage your seedlings. To learn more about horticulture, however, make up your own soil. A recipe for a general-use soil includes one unit of leaf mold (or shredded organic material such as peat moss) plus one unit of coarse sand (river sand is rounded and will pack down, and ocean sand contains killing salts) plus two units of loam (good, old dirt). Leaf mold keeps the soil fluffy with air spaces and holds onto moisture, sand helps the mix to drain extra moisture readily, and loam provides the nutrients that plants use to build tissues. If you feel that the only soil available is poor in nutrients, add a small amount of dry 10-10-10 fertilizer to your mix. Add no more than one teaspoon of fertilizer to a quart measure of soil mix. Resist inclinations to add more of a good thing. Too much fertilizer, even a little too much, will kill roots and stunt growth. Try it as an experiment, though.

WHEN THINGS GO WRONG

When nothing grows, you will wonder what went wrong. The following are some standard plant reactions to intolerable situations.

No Growth After the First Leaves Sprouted

The weather could be too cold, or something could have happened to the roots. Pull up a plant and check the roots. Healthy roots will have gleaming white tips, vigorously growing tissues that push through the soil, absorbing nutrients in water solutions. If the roots are brown or dead they could be (1) dried up from lack of water, (2) drowned from too much water, or (3) burned from too much fertilizer. You may already be suspicious that items 1 or 2 occurred, but not necessarily, as inexperienced growers will sometimes do both in succession. Too much sunlight may also tax the new roots past their ability to provide water. Start over with new resolve and, perhaps, a new location.

Leaves Turn Brown on Their Margins or Wilt

The same soil problems as 1, 2, and 3 above apply here. The roots have been damaged at a later stage and not enough water has gotten to the leaf. Wilting often accompanies margin burns. Try repotting to correct the conditions.

Leaf Yellows, Falls Off

Yellowing can occur after an intensely dry period. If there are other leaves on the plant that are still green, make a resolution to be more even-handed with the watering schedule and keep going. If many of the leaves are yellow or yellowing and watering has been regular, the problem may be lack of nitrogen in the soil. For established plants, a liquid fertilizer "plant food" with a high proportion of nitrogen (the first number of the three numbers) may help. Use the fertilizer as recommended for houseplants.

Stunted Overall Growth, Not Many Flowers or Fruit

Try a general liquid fertilizer, or one with elevated potassium and phosphate. Check the soil to see if it has become compacted with watering. If it has, lighten it with compost or peat. Use less sand in your mix.

Whenever you are trying out a new soil mix, pot and set aside several "sacrifice" plants in the mix. Treat them as you do the other plants, and check them every week or so, shaking out all the soil to see if the roots look white and healthy. If the roots stay in the upper portions of the soil, the mix might be too heavy with sand, or the watering may be wetting only the top layers. Make sure the pot is thoroughly saturated with each watering, then wait until the soil surface is dry to the touch before watering again.

98

Experiments with Growing Plants

RESPONSES TO LIGHT

A geranium will orient its leaves to face the full sunlight. If a geranium is grown on a window ledge, all its leaves will tilt away from the dark room toward the light. Seedling plants of many species will also show dramatic responses to light. Often the entire seedling will lean toward a single source of light. If turned away, the plant will reorient to the light in an hour or so. Although it is tempting to refer to the plant as "wanting" sunlight, this orientation for maximum sunlight is maintained by differential cell growth. That is, cells on the shaded side of a stalk or leaf stem elongate more than do the sunlit cells. The result is a curving of the stem and a turning of the leaves toward the light. Shade a few seedlings with little foil caps to show the effect of lack of direct sunlight.

If the plant is grown in the dark and *neither* side of the stem receives light, all the cells along the growing stem will elongate. The whole plant will be thin and weak-stemmed. Grow one set of seedlings in the sun and one set in darkness to see the difference. Plants grown from bulbs will also demonstrate this phenomenon of etiolation, or pale, weak stems when deprived of light.

By growing a plant in a box set up as a simple maze (see illustration), the photosensitivity of the stem cells can be observed. Use a naturally vining plant such as a bean seedling to avoid overstressing the plant. Grow a control plant of the same species under standard conditions. Compare the relative total lengths and also the comparative lengths of the internodes, the stem lengths between leaf attachments.

The leaves of a plant are also influenced by sunlight. The stems are

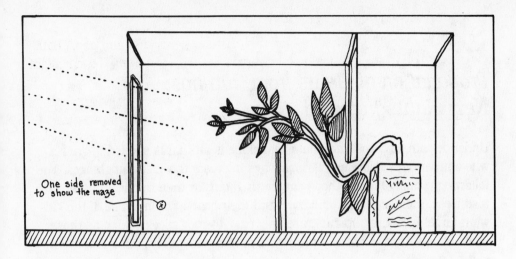

One side removed
to show the maze

responsible for the leaf orientations, but the color and size of the leaf relates
to the amount of radiation the leaf receives. By growing one specimen in
bright light and another in shade, you may be able to cause the shade-grown
plant to display a darker shade of green and larger leaf. The darker green
indicates a denser clustering of chloroplasts, the green parts of a cell that use
light to make sugars. More chloroplasts indicate a heightened ability to utilize
the less intense light. In bright sun a dark leaf would be likely to overheat and
wilt from overactive transpiration of moisture.

Block out parts of the light spectrum by growing plants under colored,
transparent films. A red transparency, for instance, will reduce the amounts
of blue and yellow light in sunlight, thereby cutting down on the energy enter-
ing the leaf. Blue would reduce reds and yellows. We know that green is not a
useful color for plants because green is the color the leaves reflect instead of
absorbing. Would a plant be able to grow at all under green film? Remember
to grow a control plant under a clear transparency to keep conditions as comparable as possible.

Even covering a part of a leaf with reflective foil gives a good demonstration of the effect of sunlight on a leaf. Reflective foil blocks *all* light from the
leaf surface. Without the stimulus of sunlight, the green pigment chlorophyll,
the coloring within the chloroplasts, does not form. Try a similar experiment
using colored transparencies on parts of a large leaf.

For Discussion

Our animal heritage of "eat to live" makes it very difficult to comprehend the
plant's ability to use sunlight energy to cook up sugars out of soil nutrients,
water, and air. Be careful not to describe photosynthesis as making food *from*
sunlight. Sunlight is the *source* of power, like electricity making a television
work. The energy is also packaged into the sugar molecule, like energy in a
battery, and transferred to places in the plant where growth will use it up
(unless we eat it first!).

99

Propagation of Plants from Cuttings: New Plants from Old

Under certain circumstances, one leaf alone is all that is needed to grow a whole new plant. There are various ways to create those circumstances. The following propagation methods can be tried out for their applicability to your resources, the age of your children, and their goals. For instance, if the kids want large numbers of plants, use only easy-to-root species such as those listed below. Try the hard-to-propagate species only after successful experiences.

The easiest plants for propagation can be readily found in greenhouses or homes. Coleus, geraniums, African violets, ivy, spider plants, begonias, tradescantias, and most of the succulent-leaved plants can be propagated using leaves or stems. In all the methods, the process is enhanced by (1) consistently moist but not soggy soil conditions; (2) warmth from below; (3) indirect sunlight, as direct sun will cause excess transpiration demands and leaves might wilt permanently; and (4) sterile potting soil.

THE POTTING MIX

The potting mix must have a balance of three attributes. It must provide air spaces because dense packing will suffocate the roots. It must hold water around the cutting, and it must also be sterile, because soil fungi will kill the vulnerable cuttings. Nutrients are not needed yet, as the cutting will be drawing on reserved nutrients in its leaves or stem for root production. Coarse sand, perlite, peat moss, or various combinations of these media are excellent. Which proportions work best for your situation is a good first experiment to try.

If you find that many seedlings are dying at their roots, you may need to sterilize the potting soil to kill the fungus. Pour boiling water through small portions of the mix, about a quart of soil per quart of water. Let the water drain through before using the soil.

CONSTANT TEMPERATURE, MOISTURE, AND LIGHT

A cutting that is put through many changes is going to give up and die. Keep conditions as moderate as possible, including heat and light. Moist conditions can be maintained around both roots and leaf to create a "greenhouse effect" by slipping a glass jar over the cutting if it is in a pot. Or you can skip the pot and put the mix and the cutting both in a plastic bag, close the top, and put it where it won't be jostled. This method lends itself nicely to large numbers of stem cuttings. Try to keep leaves away from the sides of the container. Rot will occur where tissue rests against a moisture film.

Check the illustration on page 117 for suggested sizes of cuttings to use. In general, the really small cuttings don't survive because they lack nutrient reserves, and the monsters need roots faster than they can grow them because of water demands from photosynthesis. Any question you have as to what works and what doesn't is best answered by an experiment. For instance, would a geranium stem root faster with leaves or without, or if half the leaves are cut off? Do a series of cuttings the same way and set them in various places around a room. What are the conditions that promote root growth?

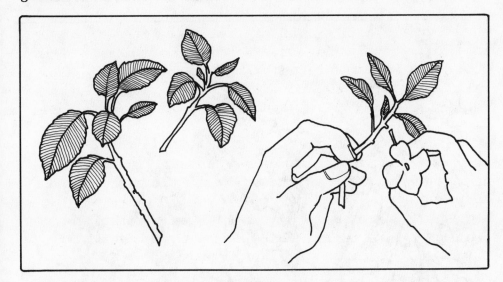

Root hormone powders can be purchased from garden centers to speed up root growth. All the plants listed on page 116 as easy species will root without hormones or with the least powerful mixture. Some plants root so easily, the hormones don't make a difference. Try samples both ways.

The cutting has rooted when a gentle tug on its top meets with resistance. It's a good idea to make many cuttings, as constant checking can be detrimental in itself. Remove the cuttings by lifting the plant from below with a knife blade or spoon and transplant them in separate pots or a flat using a standard commercial potting mix. When potting succulents, increase the amount of sand in the mix. Continue to guard against the soil drying out until the roots are firmly attached.

100

Poking into the Soil

A study of soils can be as simple as comparing the compaction of a well-worn path to the springiness of untraveled areas by bouncing on one's heels and sensing the soil's compaction through the jolts to the bones, or by pushing a pencil into various soils, using the palm to sense the differences in resistance. The differences cannot be quantitated but they can certainly be felt.

Another level of study is a sensory exploration of soils: comparing the qualities of grittiness, softness of the partially decayed leaves (or duff), moisture retention (felt as coolness to the skin) of the lower strata, or color differ-ences in soils of different areas. A highly organic, nutrient-rich, and moisture-holding soil is dark brown and consists largely of tiny leaf particles.

If there are areas in nearby woods where leaves have naturally settled and stayed, the leaves will probably show annual layering by their degree of decomposition. The topmost leaves will represent the latest accumulation. The slightly more decomposed material underneath the top layer was green

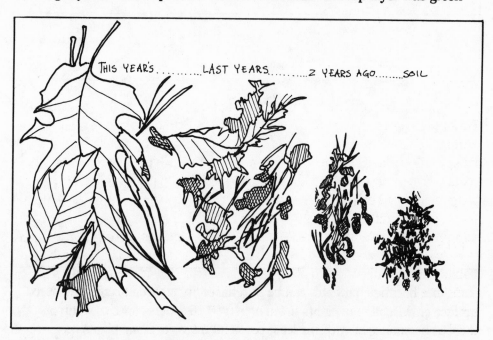

THIS YEAR'S..........LAST YEAR'S..........2 YEARS AGO........SOIL

leaves on a tree two summers ago. Notice the relative thickness of the layers. The older accumulations are usually thinner, more compacted and reduced. A third distinction of decomposition is usually the last discernable layer. Under that, the layers run together and are woven through by roots and animal tun-nels. According to one estimate, it takes five hundred years for one inch of decomposed humus to form.

Many of us have been admonished for playing in the "dirt." Some chil-dren may be hesitant to explore the pleasant qualities of leaf mold or duff. One way to ease them into handling duff is to show them that it is easily brushed from the hands, with little or no "dirt" left behind. Be ready for con-flicting opinions as reactions to the rich smells of the earth. Some people are just not prepared to appreciate the qualities of anything that looks so messy and dirty. Your attitudes and caring gestures will make a positive impression, so long as you don't press the issue.

Gathering leaf mold for an experiment is a way of increasing the chil-dren's awareness of the qualities of duff. Fill cups of equal size with sandy soil and with leaf mold. Fill two more cups of the same size with water. Slowly pour a water-filled cup into the sand cup, observing how much water it takes to fill the cup. The amount of water indicates the amount of air spaces in the soil. Repeat the experiment with the duff cup. Measure and compare the rela-tive amounts of air space in the two types of soil. Does the result fit in with expectations?

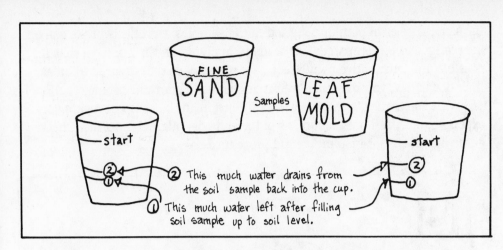

Continue the experiment by trying to pour the water in the soil cups *back* into their respective water cups. You will then have a standard for measuring the relative absorptive ability of the soils. The soil that gives up the most water is the least able to hold water available for plants to use. As the roots of a growing plant need to both breathe (exchange gases) and absorb water, what kind of soil recipe would be best for growing different kinds of plants? Recall different plants and their habitats (deserts/swamps) to keep in mind that some plants like excessive wetness or dryness.

Another method of measuring the water-absorbing capacity of various habitats is to pour water from one brim-filled can into another can of the same size that has both ends removed and that is pressed firmly against the surface of the soil in question. Measure the time it takes for the soil to absorb the water. Times can vary greatly just on one slope, or between a path and its border. Help the children distinguish between the slow absorption time that occurs with compacted soil, which would be a difficult place for most plants to grow, and the slow absorption time that occurs with a moist, humusy soil, which would be a nurturing environment.

Use different types of soils for growing seeds. Suggest that the kids scout around the neighborhood for various colors and textures of soils. If seeds of

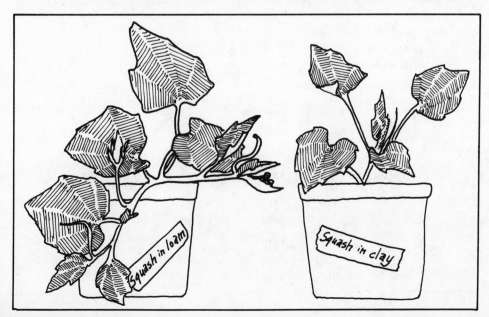

the same species of plant are started in the samples, the effects of the soil on the plants can be compared. Use seeds from plants growing in local soils if possible. As some native plants require a period of cold or dryness to break their dormancy, you may wish to use dry beans to guarantee some results.

Other activities to include in soil studies might be compost making or experiments with those champions of the compost pile, earthworms. Check gardening books for inspiration and procedures.

101

Soil as an Art Medium

We tend to think of soils as dark in color but many areas of the country have brightly colored clays that tint the earth many hues. Clays and other mineral elements used to be the main sources of color for paint for most of human decorative needs. The Early American red barn and white farmhouse were the products of clay-based paints.

Children can still make use of the colors and textures of local soils in simple art projects. Many soils will make colored smears when rubbed over stiff paper, especially if the soil is moist. These smears can be a way of "collecting" the soil samples on a walk. The sampling is especially impressive if the walk covers areas with varying soil makeup and dampness. Label each smear with the location and habitat characteristics.

Textures can be included in the samples by covering an area on a paper with white glue and then sprinkling a particular soil over it. If a number of samples are available, with contrasting colors and textures, this technique can be used to "color" a drawing with soils, creating a mosaic effect. The children may still prefer the bright colors of commercial art supplies, but they

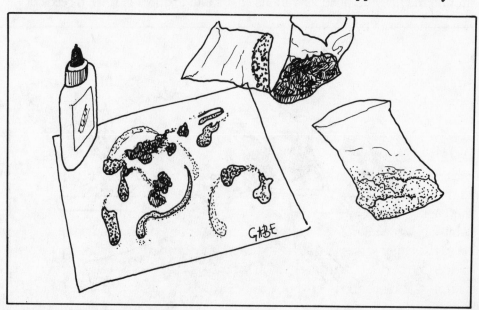

will have had a chance to handle and compare a variety of soils. Encourage them to name the distinctive soil types, and use their words later to describe other soils or textures.

102

Acid Soil or Alkaline Soil

People are becoming increasingly conscious of the damaging effect of extreme acidity in rainwater. The habitats in which it does the most damage are those where soils are already acid. The plants of acid soils are adapted to growing there—they often cannot grow in neutral soils—but increased acidity pushes them beyond their tolerance and they die. This jeopardizes the lives of the dependent animals, as well. Some acidic lake environments have been kept in balance by the addition of large amounts of lime, which brings the extreme acidity back to a tolerable level. The range of acid to alkaline is a sliding scale, largely influenced by the presence of mineral deposits. In general, the trees of coniferous forests, standing on soils derived from granite, live on and create acid soils. The soils derived from limestone are alkaline (which is also called sweet or basic). Even if the areas are in the same climatic zone, different species of plants grow in alkaline and acid soils.

An interesting project is to collect a number of samples of local soils and test them for their relative acidity. Soil kits from a garden center have special chemicals that react colorfully to the soil. A color chart shows the relative acidity in terms of pH, or "potential Hydrogen." A low number (below 7) indicates acidity, 7 is neutral, and above 7 is alkaline.

Rhododendrons and azaleas must have a soil pH of 4 to 4.5 to grow and bloom well, while cabbages need a sweet soil, with a pH of 6 to 6.5. Too alkaline a soil can be made more acid by the addition of peat or sulfur. Alkalinity can be increased by adding limestone or wood ashes.

A homemade testing solution for acid/base testing comes from a surprising source: purple (or red) cabbage. By simmering sliced red cabbage in water anthocyanin is released into the water. Once cooled, this solution can be safely used in a manner similar to the soil kit chemicals. The anthocyanin will change color relative to the acidity of the soil. Set a teaspoonful or more of the soils to be tested in a small jar, such as a baby-food jar. Plastic bags work well, also, and can be taped up against a window to observe the colors. Label each container with a masking-tape label, including a description of the source of the soil. Cover each sample with cabbage solution, wetting the soil thoroughly. Use enough solution to suspend the sample particles. Set the sample aside until the particles settle and the solution color can be seen clearly. Include some peat moss and some lime or wood ash as samples to get an idea of the full color range. The acid peat should look purple; acid soil, red; neutral soil, blue; and wood ash, green. Playing around with household chemicals such as acid vinegar and basic baking soda can also be fun, making spectrums with "magic" color changes.

103

Plant Food: Fertilizer Supplements

If you have already taken apart a large seed such as a bean and seen the small embryo with its large food packet, and if you know that plants can make their own food in their leaves, the frequently heard admonition to "feed" plants with fertilizer may seem confusing. Most children assume that the fertilizer is something like hamburger for plants. In fact, the "food" is minute quantities of mineral nutrients; it is literally the stuff of which a plant is made. Unlike our food, fertilizer contains no energy; in a plant, energy is supplied by sugars made by sun-power.

The nutrients can come from a variety of sources. It can be given to plants in the form of commercially produced fertilizers or as organic amendments such as manures and composted vegetation. Purer forms can come from ground-up minerals, bones, or ashes of other plants. Adding plant food to the soil increases the availability of nutrients for the plant roots to absorb. The soils may be naturally lacking in the nutrients or previous crops may have used the nutrients up.

Most commercial fertilizers contain varying proportions of nitrogen (essential for protein making and expansive growth), potash (important for stalk strength and fruit production), and phosporic acid (which makes strong roots and abundant fruits). On any given package of fertilizer, the proportions of these mineral nutrients are noted in their relative proportions. For instance, a fertilizer high in nitrogen, probably from animal waste products, would be labeled 10-0-0. A general fertilizer would be labeled 10-10-10.

You can easily demonstrate the effects of the different elements by growing several plants in varying proportions of nutrients. Use a single kind of seed, perhaps a grass or flower seed instead of beans, because special bacteria will fix nitrogen in the presence of bean roots, throwing off the experiment. The potting mix should be a nonnutritive medium such as sand or perlite, so that whatever plant growth occurs will be a reflection of the kind of fertilizer added to the mix. You won't be able to compare the results, of course, unless one set of plants is grown as a control, without any fertilizer. Use the same amounts of water on all the samples. Try as many proportions and as many brands as you can find or afford. Follow the package directions for mixing.

Keep track of the plant growth over time, as some fertilizers may trigger rapid leaf growth, then not help at all when the plant is ready to flower and fruit. End the experiment for all plants simultaneously and measure their root growth and their weights, both while the plants are fresh and when they are dry.

For younger and/or less patient children, an effective experiment can be set up using grass seed and two new sponges. Sprinkle both moistened sponges with equal amounts of lawn seed, then keep the sponges moist with equal amounts of water, adding liquid or dissolved fertilizer to the water of one of the sponge lawns. Compare the grass blades as they grow, noting the size of individual leaves, their color, and their relative strengths, or their ability to hold themselves upright. Another variation is to use a pot of regular soil mix with the same kinds of seeds and the same amount of water, but add some fertilizer only to one.

For Discussion

Plants that grow wild don't get bottles of plant food poured on them. Where do the nutrients come from in nature? Nutrients are tiny pieces of soil and molecules from the air. Just as it might be difficult for us to get nourishment from a whole pig or corn plant, the soil and air particles need to be made small enough and processed for a plant to use. The best form of fertilizer for most plants to use is the recently decayed structures of other plants. Most commercial fertilizers are products of petroleum or coal industries. Where do oil and coal originate?

104

Open Terrariums

An open terrarium is more like a habitat than is a single potted plant. A group of plants that have similar soil and sun requirements can be planted together in a deep pan or tray along with other interesting "accessories," such as rocks or wood. Instead of centering the plants in the middle of the pots as is done for single plants, position the plants for their relative needs and for their aesthetic appeal. The results are invariably interesting. Even if each child is given the same plants and the same-size container, each finished product will be unique.

The requirements for terrarium soils are similar to those used in standard houseplant mixes. Use a commercial potting soil, or mix together two parts loam, one part peat, and one part coarse sand. A teaspoon of dry fertilizer per quart of mix will keep the plants growing. Ideally the container should have holes in the bottom for drainage. If not, lay at least an inch of broken clay pot pieces on the bottom. Broken-up bits of foam coffee cups will do in a pinch. Cover the pieces with a layer of peat to keep the soil from washing down and packing the drainage spaces. A cloth with a very open weave, such as cheesecloth, will also help maintain the drainage when laid over the broken pieces.

Then fill in the soil mix to the desired levels. With a tray, try to spread the soil so that there are some little hills and valleys to give an interesting aspect. A major feature, such as a rock or a piece of wood, will add visual interest to the planting. Plant the larger plants first, to establish the major focal points, then set in the smaller ones. Let each "landscaper" make decisions on placement without your help. Insist on children's personal preferences as the criteria for the best placements. Leave time for changes and alterations.

The life span of each habitat depends on the choices of the plant species and the care given them. It is important that the species have similar requirements: tropical houseplants that like shade should not be put in with the sun-loving succulents. A desert plant garden is an easy and excellent start for young horticulturists. Many cacti and succulents are thornless. All have interesting shapes. They also won't grow fast enough to disrupt the original

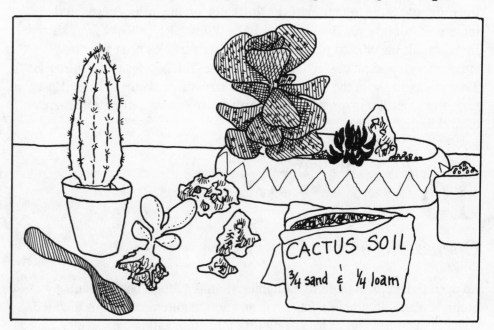

spacing. They do need sun and a soil that drains readily. Add one more part of sand to the mix, as well as some rough pieces of gravel or brick to ensure drainage. Use sand for a soil covering to continue the desert theme. Remember that desert plants do need some water: weekly watering during the summer, at least monthly during the winter.

105

The Effects of Sunlight or Soil on a Plant Species

The easiest way to determine the effects of sunlight or soil on a plant species is to investigate the various patches of a plant common in your area. Broad-leaved plantains, dandelions, or other weedy species grow in a variety of environments near where humans live.

Take a walk around outside and then make a list of variables that might influence the plant growth. A possible list might include: sunlight, moisture in the soil, compaction of the soil, amount of foot traffic on the plant, or competition with neighboring plants.

Whatever plant the children choose to study will demonstrate various tactics or responses to those variables. Collect leaves (either directly or by outlining on paper) from the contrasting habitats and compare to see if direct correlations can be made from the size of leaf, length of stem, number of flowers or seeds, color of the leaf, or amount of insect damage to the plant. (Insects seem to do more damage to stressed plants.) Devise other means of measuring the variables in the habitat. Some may be subjective, such as feeling the coolness of the soil to measure the moisture content or the hardness of the soil by making a stiff-legged stomp on the earth to sense the impact on one's own joints.

If you have access to a field or yard that is shadowed by a building or a large tree, the lack of sunlight at different times of day may affect the growth of plants in the area. Leaf collections may indicate that plants of the same species grow more sparsely or have larger leaves in the area of midday shadow, or you may find that entirely different species live in the shade. Mark the dimensions of the shadow at several times during a sunny day in late spring or summer and see if the changes in plants correlate to the outlines of the shadows.

Encourage brainstorming for other ways to compare areas. Sunlight causes fading in some materials, as well as evaporation and dehydration. Whichever means of comparison you use, try to test for one variable at a time

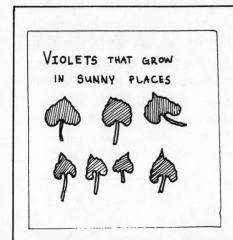
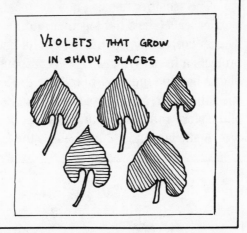

by keeping other factors equal. For instance, when counting plants in shady and sunny areas, compare plots of the same size.

Whatever the results, have craft material handy to stimulate the creation of vivid charts or diagrams to express the results of the experiments. Set aside time for explanations and discussions of further experiments.

106

Body Language in Plants

As the children do experiments that produce contrasts between healthy and stressed plants, help them evolve a vocabulary and a sensitivity to general and specific attributes that distinguish states of health. Many of the attributes are easily described: colors, textures, sizes, and positions of leaves. But some plant "gestalts" can be expressed by alluding to something analogous. The similar form or situation could be inanimate ("like a broken toy") or animate, personal ("like a painful sunburn"). Initiate a game in which various plant debilitations are mimed, or the children might draw pictures of themselves as plants, complete with appropriate colors. Illustrate "befores" and "afters," showing plants with all they need for health and then the same plants under stress (or vice versa).

107

Plants that Move on Their Own

Plants that respond visibly to external stimulation are a source of fascination. We have grown so used to plants behaving stoically around us that when a plant flinches or otherwise responds, it is a thing of great interest. It therefore is a very good tool for teaching about plant life in general.

The sensitive mimosa has the most striking reaction to touch. Its many slender leaflets fold flat together and the entire leaf bends slightly. This plant lives in hot, sandy, open habitats. Its ability to respond to touch or to wind protects it from drying winds. Its response may even startle timid herbivores. Plants have no muscles, of course; the leaves are closed by the swelling of special cells on one side of the stem. The uneven stretch results in a bending of the stem and the shutting together of the leaflets.

A similar kind of hinge closes the fatal leaves of the Venus flytrap. The

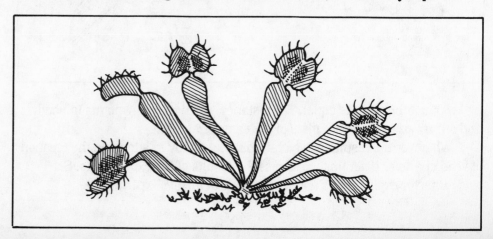

trap is sprung only if special hairs on the pads of the open leaves are touched more than once, in the manner of a moving insect. A single touch, from the chance movement of a grass blade, will not trigger the trap to close.

Some seedpods will respond to touch by bursting open and scattering their seeds. Ripe violet pods, jewelweeds and their garden relatives, and Patient Lucy all have seed pods that pop open in a startling way.

Other plants close their leaves at night, thereby reducing evaporation or damage by nighttime foragers. Tree mimosas "go to sleep" at dusk. More convenient for indoor study are small plants that can be grown in pots. White Clover, oxalis, and bush beans can be induced to close by placing them in darkness for an hour.

Some of the spring-flowering bulbs, such as tulips and crocus, will open and close in response to temperature changes. A cool temperature causes their petals to close, and they open up when brought into a warm room.

The seeds of Broom Grass (*Andropogon scoparius*) have a tiny "weather watch" attached to them. When moistened, the tightly coiled spring unwinds in a corkscrew gesture. As it dries, it rewinds, but too slowly to be easily observed. This winding and unwinding may help bury the lightweight seed in open soil.

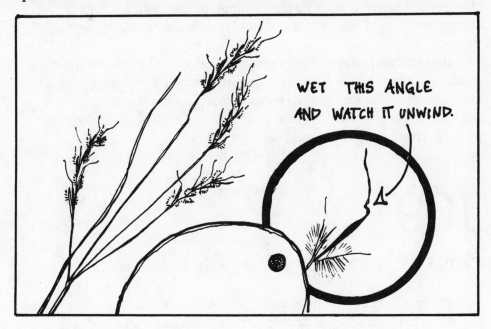

WET THIS ANGLE
AND WATCH IT UNWIND.

108

Row Plantings as Environmental Indicators

Humans are much given to setting plants out in rows, either to facilitate agricultural practices or to define the boundary of a territory. A row of plants of the same species immediately becomes a means of evaluating incremental differences of soil, sun, moisture, or pollution.

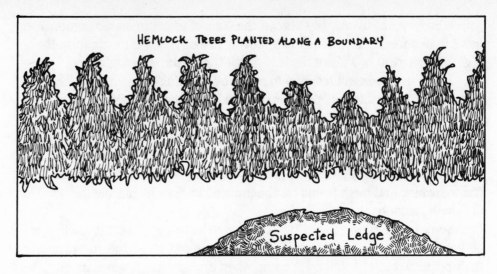

HEMLOCK TREES PLANTED ALONG A BOUNDARY

Suspected Ledge

For instance, any crop rows will dramatically show where the soil was too wet for sprouting, too dry for sustained growth, or where fertilizer was spread unevenly. Rows of evergreens, planted for hedges, can indicate pollution damage from road salt, stress from car exhaust, shallow soils, or poisoning from dog urine. A line of deciduous trees along the length of a road may show uneven patterns of spring and autumn transformation, especially around a streetlight, which may confound the photosensitivity of nearby trees.

Have children collect descriptions and/or photos of examples of responses among a row of plants. Create a miniature example by planting a line of marigolds or similar fast-growing annuals in a planter or a garden so that some are partially shaded, or place a flat rock under the roots of several plants to create shallow soil conditions.

109

Plants as Indicators of Roadside Pollution

Modern roads are stressful habitats for plants. Pavement blocks water and air to the soil, road salt and petroleum products poison the water, and lack of shelter exposes a plant to the full brunt of weather extremes. Trees planted on the edge of pavement must get their moisture from the unpaved areas. As the roots grow in their search for water, they may bend around and grow so close against the trunk that they strangle one side of the tree, killing the living tissues of the bark. Road conditions also seem to reduce the plant's resistance to pests. Tree limbs over roads often show increased numbers of deformities caused by insect and virus infections. Leaves of plants near roads may be brown and dry, especially the evergreens with their year-round leaves. One kind of pollution blocks the formation of chlorophyll in new leaves, so that the plants have white leaves. In autumn, deciduous trees that are stressed by road conditions will be the first to turn fall colors.

One group of plants, the lichens, are known as pollution indicators. They will not grow near heavily traveled roads or in areas where the air pollution is high. These are tough plants, able to grow on barren soils or trunks of trees. While we may not miss them (they have neither glossy leaves nor flowers), their inability to grow in polluted environments should be taken as a warning and reminder of our own vulnerability.

Have children collect examples or arrange a trip during which sample counts can be taken of the variety of plants in the cities compared with the variety of plants in less populated areas. Make drawings or take pictures that show the conditions of the plants, especially the longer-lived trees.

Compare these with specimens of the same species in healthier environments.

110

Charts and Graphs: Facts at a Glance

If you were to plant five beans in five different soil types and water each pot the same amount to observe the effect of soil on bean growth, at any time during their growth, the beans would be a living "chart" illustrating the effects of the soils. After your experiment you might want to make the information permanent by pasting strips of paper, cut to the actual sizes of the plant, next to each other on a line. The plants could also be represented by other than the actual units of measure. For instance, if beans are five and eight inches high, the symbols could be five and eight *feet* high. However the information is represented, visual data quickly reveal which plants are tallest, smallest, or of equal size. The graphic qualities of a chart give the brain a chance to perform one of its best tricks: computation of related material into hierarchies. The hierarchy can be of the most-to-least variety, such as variously grown beans, or can relate to changes in time, such as a chart of drawings that show daily growth of a plant for a month.

The collecting and recording of data is seldom as interesting to children as is the presentation and discussion of the results. Cause and effect and the concept of multiple variables is beyond the interest of younger children (those under eight, usually). For them, a stick on which sequential growth is measured, especially if it is rapid growth (use corn, ailanthus trees, vines, and so forth), is more understandable. Young children have flexible and uneven notions of time, and the concepts should not be forced upon them if they are not ready. When they *are* ready, they may well remember the experience of measuring plant growth and be able to relate it to the abstract idea.

The following are some ideas for simple presentations of data in chart form:

■ Bar graphs—Use bright, contrasting colors to highlight distinctions.

■ Lines plotting peak measurements—Connected points can show the rate of changes in growth over time.

■ Expanded or reduced proportions—Enlarge the chart for dramatic emphasis. The translation of the proportions is also a good math problem.

■ Extrapolated hypothesis—"If a bean plant grows this far in one month, how much taller will it be in two weeks?"

Measure fast-growing herbaceous plants, which will grow the size of a child over a period of weeks. Try perennials such as ferns, hollyhocks, lilies, goldenrods, or weedy annuals such as sunflowers, tomatoes, and garden vines. Measure the child at the same time and record the height in the same manner. Young children can show growth themselves in a month's time.

111

Plant Math: Growth Curves and Averaging Data

When two plants are grown under different conditions, the distinctions between them may be easily observed, and measurements may seem unnecessary. However, measurements that are recorded as a plant grows can help uncover relationships and patterns that would otherwise escape notice.

When tracking the growth of a plant over time, a bar graph or a plotted curve can help you see relative rates of growth. When combined with other information, such as the number of leaves at the times the measurements were made or the flowering or fruiting times, a pattern begins to appear.

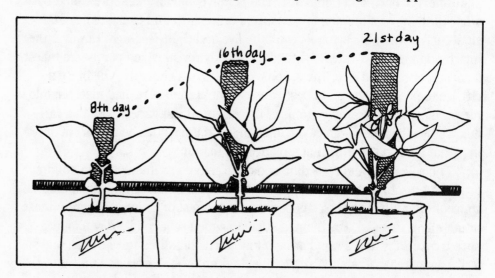

Other measurements can put greater detail or relief on the comparisons. Individual leaves can be measured, flowers and fruit capacity counted. The internodes or stem spaces between leaf attachment sites (called nodes) vary with the growing conditions. When compiling large samples of measurements of the same structures or events, it is useful to average the information by totaling all the measurements and dividing by the number of samples. In this way, for instance, an average size of an internode on a sandgrown bean may turn out to be two inches, compared to a four-inch average internode on the control plant that is grown in a balanced soil mix.

As with any other presentation of experimental results, encourage the children by examples and by suggesting activities that lend themselves to graphic display. Provide colorful and unusual materials to challenge people's resources. Grasping the implications of environmental relationships is more important than memorizing specific measurements. Make it a habit to challenge the conclusions by questioning children as to implications and further possible experiments to test the assumption. Try to keep it light. Inquiry should look like new doors, not brick walls.

112

Activities to Practice the Art of Careful Looking: Transects and Quadrats

All the scientific ideas that we accept as "facts" are the result of two hardworking procedures. The first procedure involves making a guess at what is going on (a hypothesis), and the second requires that the proof of the guess be uncontaminated by the hopes of the guesser. The guesser must prove that many samples have been reviewed and that the samples as a whole represent the way nature works. Both procedures are based on the idea that an understanding of natural processes must come from carefully looking at (1) what is there compared with (2) what we *think* is there. The second item may well cause us to miss some important factor and a new, more useful understanding.

Collecting a large amount of data slows us down; we are more likely to notice a significant attribute when we have to make subtle distinctions. A transect is a tool to get us to notice changes over a range of ground. A number of transects will give us ideas about how the elements within the given area change or form a pattern. Here is how it works: Run lines of equally spaced string up a slope or out from the edge of a pond. Mark numbered units along their lengths to facilitate cross-referencing. List all the plants found growing next to the string and then correlate the information graphically. Thus the children will get an overview of the effect of habitat change on plant species.

If the change in plant species is very clear, it may not be necessary to know the names of all the plants. Drawings of their leaves and habit of growth may give good indications of their adaptive responses. For instance, wetland species are more likely to have thick and juicy leaves, while dryland species may show modifications for drought-resistance, such as narrow or hairy or light-color leaves. (This is a good hypothesis to test with a transect!)

A transect helps you compare changes in a line. A quadrat measures in two dimensions. A quadrat is a tool for looking at and comparing equal-size plots of plant groups. If the quadrat is small and mobile (a stiff hoop or a sturdy square or rectangle), it can lend itself to random selection of sites, the heart and goal of all data-sampling processes. Instead of locating and counting all the plants of a field, a small quadrat lets a scientist take a set number of random samples, then compare them with the same number of random

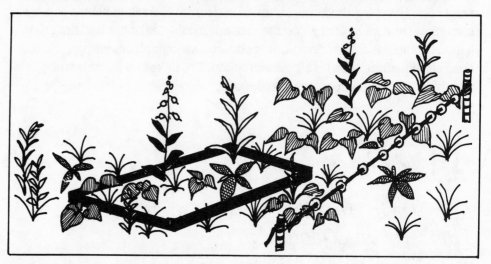

samples taken in another area. But how to make it random? How can you select "random samples" without influencing the outcome? The standard scientific method is to toss the quadrat over your shoulder in as random a fashion as you can muster. It is important to repeat your tossing technique with every sample.

Particularly if you are working with older children (third-grade and up), encourage discussion about the methodology of data collection. Help kids to see (if they don't come upon the notion themselves) that standard units and random sampling are essential to the process of objectively perceiving patterns and processes. You yourself will observe the world in greater detail through this process.

113

Background Activities for Understanding Plant Succession

It is fairly easy to get birdseed to sprout and grow in a pot of soil, but it is almost impossible to get a second sowing of the same seed to grow through the crowded stems of established seedlings. Larger-seeded plants are more successful—sunflower seeds, acorns, or beans can usually push their way above the grassy birdseed plants. On a very small scale, this sequencing of plant growth, from bare soil to quick weedy growth to larger, slower-growing plants, is the basic demonstration of plant succession.

Plan a walk outdoors to find examples of succession. Any newly disturbed soil should show some plant growth typical to the aggressive first settlers: small-seeded species that grow quickly *up* (grasses) or spread quickly *out* (dandelions or plantain). Thicket-forming species, plants able to put on sprawling, woody growth, are the next to follow. In their shade, the larger-seeded, slower-growing tree species will be getting their start. When you investigate a woodland area, look at the undergrowth. Usually the undergrowing plants will be of a different, more shade-tolerant species than the trees overhead. At some point, the succession will peak and one group or species of plants will predominate. The story is not over, because unusually severe weather, disease, or old age will alter the system, and parts of the woodland will revert to "younger" levels of succession. Experiment with the concepts just discussed on a minihabitat. If a larger plant is pulled out of a weedy spot or a garden, do the smaller plants respond over time to the opened space?

114

Developing an Awareness for a Habitat: Ways of Describing and Giving It Value

Assign various habitats or otherwise distinct areas to individuals or small groups. The object of this activity is to describe their spot in such a way that others will want to "visit" and see for themselves. Travel brochures and posters from a travel bureau will help set the tone for fun and alluring language. Encourage children to use playful hyperbole. Have materials for poster making available. Even though the language may become exaggerated, both the salesperson and the tourists will get a new point of view on an area. Leave time to discuss what qualities make a piece of land special. Ask if having to write about a place made a difference in the writer's perceptions.

Other variations on the theme of travel inducements: which attributes would be emphasized to sound attractive to a blind person, a land developer, an artist, or a small child?

115

Seasonal Changes in a Special Plant

Children will become personally involved if they can choose a particular plant to watch and then report on its changes. They may want to know the common name for their plant, or they may wish to make up an original name. The method of reporting can be varied for interest: drawings, descriptions, rubbings, collections of spare parts, lists of visitors, interpretations of the plants' reactions to different events, life histories (real or imagined); all can play a part in helping children become more aware of special plants in their lives.

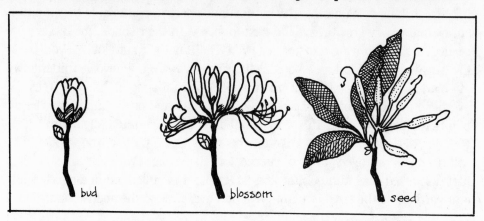

bud   blossom   seed

The adult's role is to set aside special times for visits to the plant and then to listen to the reports. Help a child create records of the experience that will mean the most to him or her. Allowing the children to use special equipment, such as tape recorders, or taking a photo of each child and plant, will help the children give value to their work and their plant.

DIARIES

Natural events can sometimes seem like personal messages or metaphors for our lives. For older children, keeping a diary of regular observations of a special plant, with encouragement to compare their observations with their feelings, invites both keener observations and deeper connections with nature. In discussing nature literature and diaries, it is important to help the children distinguish between the useful reflections of nature as analogy (using a tree's life to understand your own) and the useless anthropomorphosizing of nature (using your life to understand a tree). We do get feelings and insights from nature, and they can be meaningful, even crucial to us. Yet it can be misleading to impose on plants our hopes and goals or our feelings of success or defeat.

Gently guide preschool and newly literate children to see the difference between "the tree felt sad when" and "if I were a tree, I'd feel sad when . . ." The latter is a good starting place for understanding nature. Maybe the only one.

116

Dandelions from Top to Bottom

The ubiquitous dandelion is often literally around our doorsteps along our paths. Dug out, poisoned, and even eaten, it remains fabulously successful. It must be quite remarkable to survive so well.

The following are aspects of the dandelion that make it worthwhile as a subject for study. You could easily spend a year learning about this plant. Try to set up the study so that the children themselves discover as much of this information as possible.

■ The dandelion flower is typical of other members of the composite (Daisy) family in that its yellow face is made up of hundreds of individual flowers. That is, each tiny flower has a seed, a stigma, stamens, and petals. These parts are so modified that they are hard to recognize. (See illustration.) The modifications make it possible for the many flowers to fit together in a small

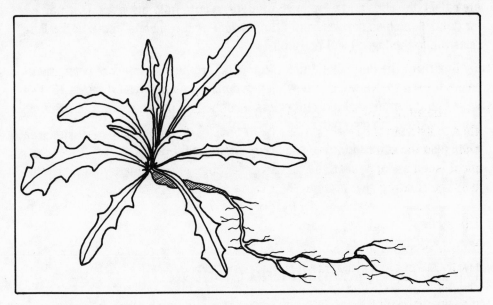

space. The arrangement of the flowers makes a large, attractive landing field for an insect. As the insect walks over the flowers, many seeds are fertilized by pollen that falls from the visitor's hairy underside.

■ For how long does a dandelion bloom? Under what conditions does the flower head open? Is it open on a rainy day? How many flower heads are on one plant? Does the number indicate anything about the plant's health?

■ Once the flower head has finished blooming, the bracts (the fringe of bud covering) close again and the stem bends in such a way that the bud is held close to the ground while the seeds ripen. When the seeds are ripe, the stem extends and grows longer than before, so that the fluffy seed head is held high. What is the advantage of hiding the seeds while they ripen? How does it help the seeds to be held up high when they are ripe? Under what conditions does a seed head open? As the seeds are spread by the wind, where would the most seeds be expected to accumulate? Do the seeds have any way to snag onto the ground once they touch down?

■ Try planting your own dandelions from seeds. Keep track of how long they take to sprout, whether the seeds all sprout at the same time, and what conditions stimulate sprouting. Have a contest to create conditions to grow the biggest dandelions *and* the smallest.

■ When dandelions first show in the spring, dig up some roots and observe their size and proportions. The energy stored in the thick, central taproot was made the summer before. Dandelions are notorious for supposedly being able to grow again from a small portion of taproot. Is this true? Is the taproot smaller after the plant flowers and seeds?

■ Is the milky sap found in all parts of the plant? Touch it to your tongue. How might that taste protect the plant? People sometimes eat dandelion greens in the early spring before the flowers form. Try eating the leaves (steam them) before and after flowering for comparison. (And for a healthful dose of vitamins.) Don't eat *too* many. In France the country name for dandelions is *pis-en-lis*, or pee-in-bed. Medicinal plants that increase the urine flow are called diuretics and were once thought to cure mild diseases. Our name for dandelions also derives from France: *dent de lion* means "tooth of the lion," referring to the jagged leaf margins.

■ All parts of the dandelion (including roots) produce a variety of permanent colors if used to dye wool. Follow instructions for plant dyes on pages 15-17. Leaves and flowers make bright colors if rubbed on paper. Color a portrait of the plant using its own parts.

■ See also the games for airborne seeds on pages 63-64.

117

Winter Rosettes: Wild and Cultivated

During the winter or in early spring a wide variety of nonwoody plants can be found with leaves arranged in rosettes. The leaves of various sizes grow about a short central stalk to form a "rose" of leaves. The rosette is budlike in that the leaves are arranged spirally, with the outer leaves sheltering the central bud.

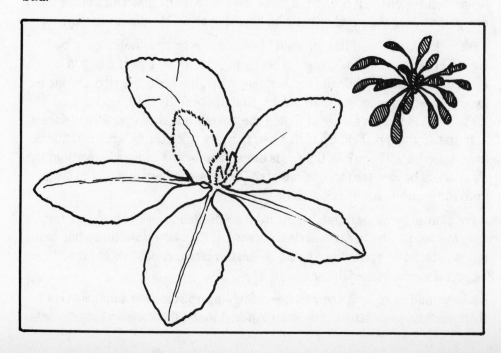

Rosettes are common forms in plants that specialize in populating recently disturbed soils. During the first summer, the plants grow leaves that overwinter in the sturdy rosettes. Thick taproots store food below the leaves. The next summer, a tall stalk grows above the leaves on which flowers and then seeds are borne. The plant usually then dies. Plants which follow this growth rhythm of blooming the second year, then dying, are termed biennial. Carrots, parsnips, and turnips are biennials that have been cultivated for their fleshy and nutritious storage roots. Endive is a biennial that is grown for its succulent second-year leaves.

Collect a number of biennials from a vacant lot or edge of a schoolyard. Whenever possible, ask permission before digging. You are not likely to be turned down since you will be collecting plants that are commonly thought of as weeds: dandelion, Queen Anne's-lace, mullein, mustard, Chicory, and Evening Primrose. All of these plants are biennials, but not all rosettes are. If the rosette comes with a long, thick taproot, however, it probably is biennial. Without their flowers, the rosettes can be difficult to identify, but it isn't necessary to know their names to enjoy their beautiful symmetry and flowerlike arrangements of leaves.

118

Outdoors in the Rain

This activity was inspired by a book by Peter Spier entitled *Rain*. There is no written text to the book. Each page illustrates facets of the experiences to be had walking and playing in the rain, including the sunshine at the end. The book will inspire you and your children to get out in the rain. Hopefully, you will also want to create your own book about your experiences. If you can't find Spier's book, get out there, anyway.

As you begin, ask the kids to describe their previous experiences and observations as to what happens outdoors during the rain. Find out if they have any ideas about the effect of rain on plants. If they are already familiar with places or particular plants, ask them to guess how they might be different in the rain. Ask them to imagine and describe the way a person or a plant might feel before, during, and after a rain. Your enjoyment of their answers will encourage a greater variety of responses and build up a vocabulary for making observations when you go out. Decide as a group (if possible) which areas you will check out, but leave time to discover new stops. Tell the children that they will be making a book when they return to illustrate what they saw in the rain. They will need to look carefully, to take "memory photographs" in their mind so that they can make a good picture when they get back inside.

You might consider using large pieces of construction paper (all the same size) for individual pictures. Filling a large page with a drawing or painting is more likely to elicit larger body movements and more dramatic expressions of the children's memories. Staple or sew the pages together and put the book on prominent display. Refer to the book on other occasions to enhance its value.

119

Plants as Historical Indicators

Most states and many towns have trees that are protected as memorials to historical events, such as charter-signing, or homesite markers of famous persons. Throughout human history, distinctive trees or rocks must have served as meeting places for special gatherings. Large trees seem important in themselves. One has only to stand under a large tree and look up to feel its dignified presence. Use libraries and historical societies to find the location of historic plants near you that could be visited.

Plant communities can also provide a historical narrative. As plant succession often follows somewhat predictable patterns in an area, the relative sizes of some plants can serve as clues to the land's recent history. For instance, within a woodland, a stone wall may divide slightly different kinds of trees or obviously different ages of trees. Larger pines on one side of the wall and much younger pines on the other may indicate a relatively recent field reversion in the land with younger trees. With woods, straggly shrubs or extremely slender field tree species, such as gray birch, aspen, juniper, or ash, also indicate an old field habitat that is being overgrown by a mature forest. Once the tree community gets growing, most of the additional growth occurs only at the tops of plants, as they compete to stay in the sunlight. A large tree with wide-spreading branches is a sure sign that the tree once stood alone in a clearing, left on purpose, perhaps, as shade for grazing animals.

120

Creating Imaginary Plants to Fit Special Environments

The goal of this activity is to make up a new plant, using material found in a particular habitat. The plant should illustrate special adaptations to that habitat. Each person might work alone or in small groups, where talents and ideas could be pooled. As you describe the activity, discuss the factors influencing the forms of plants the kids already know, or refer to some examples at hand. Include the plant's need to conserve water, absorb sunlight, attract pollinators, deter predators, and disperse its seeds. Using only materials found in an actual habitat increases the challenge as well as the awareness of that habitat.

If you can't get outside, the habitat of a particular area in a building works just as well. (A refrigerator plant? A closet plant?) The very impossibility of a real plant growing wild in a room can release the imagination from the limitations of "getting it right."

Create plants that could survive specific catastrophes. What would a plant look like that regularly survived floods, or hungry hordes of insects or birds, or lived through fires or droughts? How about a wind-resistant plant?

121

Creating Animals to Go with the Plants

As most people are more familiar with the concepts of animal adaptation, the children may find it easier to create an imaginary creature out of natural materials from a particular habitat. The material that has been gathered from a specific area can serve to camouflage the animal when it is hidden in that area. (If the creature is too well hidden, how will it communicate with others of its kind to find a mate and raise its young?) What foods and shelter does the environment provide, and what seasonal changes in body or behavior will be necessary?

As children create plants or animals, support them in their wildest flights of fancy. If they seem uncomfortable at starting, do one example as a group to indicate the procedure and how questions can stimulate creative solutions.

Bring inside a variety of sticks, leaves, and other "habitat ingredients" to create a miniature habitat in a tray of soil, sand, and/or gravel. Plan for areas to collect moisture and have places for animals to hide. Then design plants and creatures to live in the new habitat. Use leftover materials as possible "home improvements" for the new creatures. Have handy also extra materials such as cork, cardboard, toothpicks, pipe cleaners, thumbtacks, twist ties, and so forth.

122

Insects on Plants

Begin with the simple challenge to find one summer leaf that hasn't been nibbled at or altered by some insect. The munching begins as soon as the leaves are out in the spring. If a plant is significantly unblemished, check for some anti-insect device such as hairs, waxy covering, or aromatic oils.

As children discover more evidence of insect work, they should start to notice that there is variety in the styles of insect usage. Many insects use a leaf in only one way. There's the disappeared leaf, consumed by the mouthful by a hungry larva of moth, butterfly, or sawfly (a larva of a wasp). Perforated leaves, where sucking insects have deformed or darkened areas, are often the work of aphids or the true bug clan. A rolled leaf, stitched shut with silk, may house a caterpillar or sometimes a spider.

Wilted leaves or vines may be host to borers living within the stems. Look for a moth or a beetle larva. Wiggly, light-color lines or blotches on the leaf indicate miners at work, consuming the green cells of a leaf between the transparent skins of the upper and lower surfaces. Miners are most often the larvae of minute moths. Strange bumps and spheres are the most fascinating of all. You might enjoy getting a book on the variety of galls there are around us. A gall is a plant's reaction to the presence of the larvae of certain wasps, flies, aphids, or moths. Each species of insect specializes in a specific plant (or alternates between two!). The female insect lays an egg on or near the host

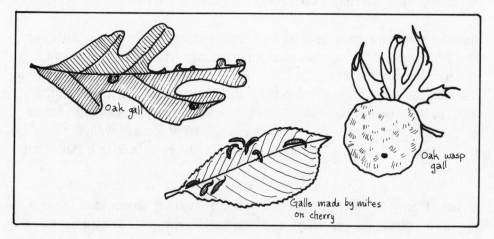

plant, and as the just-hatched young eats the plant, it releases chemicals that mimic the plant's growth hormones and stimulates the formation of a gall. The gall becomes the parasite's food and shelter. If the gall is opened carefully, a single or a cluster of tiny grubs can be found. (If you do this, you are of course destroying them.)

123

Investigating a Rotting Log

A rotting log is the last chapter in the life story of a tree, but it is the first chapter in the lives of many other plants and animals. As the wood absorbs moisture and is broken down by the organisms of decay, it will pass through many subtle and overlapping stages. Each stage is represented by a cast of characters that use the resources of the tree in a particular way. Each organism changes the tree and prepares the way for the next series of occupants. The rotting log is home and nursery for beetles, worms, crustaceans, centipedes, millipedes, and salamanders. Fungi and various molds help reduce the hard wood to spongy fibers so that, eventually, the wood becomes peaty and moist, the perfect seedbed for a variety of plants.

Investigate a log where it lies and you will limit the amount of havoc you wreak on the lives of the inhabitants. Bring inside portions for more detailed observations. In winter, especially, the warmth of a room will bring to life a great number of creatures that hibernate in rotten wood. Try to return as many of them as possible to appropriate replacement logs.

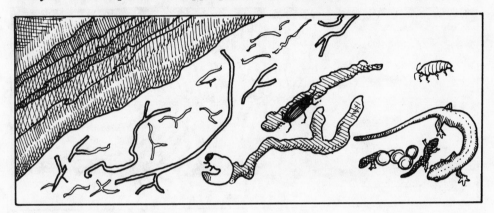

TREES: LIVING AND DEAD

Compare live and dead trees in your area. Choose trees of the same species and similar sizes, if possible. Look at the differences in bark conditions, number of branches, number of plants underneath the trees, and amount of "sky-space" each takes up. If time and patience allow, also compare the populations of animals that make use of the respective trees.

124

Beetle Writing Under Bark

The children may have already noticed the shallow, wiggly tunnels in the wood of dead trees just under the bark or on the surface of barkless wood. Many different species of beetles are born and spend their early, wormlike larval stages eating their way through dead wood. Eventually the larva stops eating, makes a special chamber for itself, and changes into a resting, or pupal, stage. On hatching from the resting stage, called a pupa, a beetle finally looks like a beetle: two stiff wings folded down to cover a pair of cellophane-like wings. In general, bark beetles are flatter than most beetles, but the small ones may be shaped like bullets. Small holes in the bark are exit holes for the adult beetles—a good indication of beetle tunnels below. Woodpecker work on a tree is another clue. The woodpeckers can hear the clamping of the tiny jaws at work and chisel through the bark after the larvae. Never fault woodpeckers for chopping into a tree. They dig after the beetles that are there because the wood is already dead.

Rubbings using crayons is a good way to collect various examples of beetle handiwork. Use the flat side of a crayon to rub color on a medium-stiff paper held firmly over the exposed tunnels. Each beetle species makes a gallery of a different design. Some wander randomly, others fan out from their birth spot to create a radiating pattern. In each tunnel, you can make out the tiny beginning and the end, the chamber of the pupal transformation.

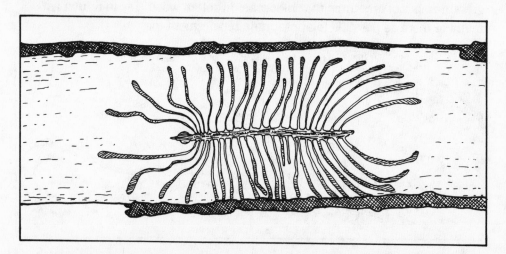

After looking carefully at beetle histories as told by the tracks in the wood, pass out paper (which, after all, is recycled wood pulp) and markers and suggest the children pretend to be bark beetles. The first dot on the page is the beginning, the egg, and the lines that go from there are the stories of their lives up to now. They can pretend to be beetles in wood (what would it feel like, eating away in the dark, hearing the scratching of a woodpecker, biting into a knot?).

125

Bird Nests

Birds use specific plants in their territories as material for their nests. Different species in the same area will choose different kinds of materials. The species of nest-builder can often be identified from the materials used in the nest. The best book for identification is in the Peterson Series of Nature Guides, *A Field Guide to Birds' Nests* by Hal H. Harrison. There are editions for both eastern and western birds.

Autumn is the *only time* to do this exercise, as birds won't be disturbed then and nests are never used again. An old nest would be weakened by the winter weather and would be more likely to harbor parasites. The tiny mites likely to be left crawling about on the nest can do no more than tickle human skin.

As the children carefully take a nest apart, have them sort out the building materials into separate piles. Some stiff material is put down first as support, then a finer material is used for "modeling," and then the finest, smoothest materials are used as lining.

Try to find the plant species growing nearby that the bird used in nest making. As birds tend to use only dry matter (why do they?), the children are most likely to find examples of spring forms of the plants.

Challenge the children to make nests similar to those they have investigated. What kinds of appendages would be helpful in making a nest as well as a bird does?

126

Who Eats Whom
(with a Plug for Endangered Species)

Even a partial listing of the food we eat (or the food of the food we eat) clearly shows our complete dependence on plant life for our existence. It is not so obvious to see that *all* life forms are connected by the passing of plant-made food from organism to organism. When investigating a particular plant in its habitat, look for plant consumers and then try to figure out what creatures would eat the plant eaters, and what *their* predators would be. Eventually you will reach some top predator that is in turn held in relatively low numbers because of its difficulty in catching its prey or because it needs large territories to secure enough prey. When children are reasonably familiar with a number of local food webs and can visualize the members of the web, use a sitting-together time for the following activity.

This first part is optional, but it adds to the degree of identification the participants feel with their organism. Pin the names of organisms known to all the children on their backs, one per child. The players must then find out what names are on their backs by asking questions that can be answered only by "yes" or "no." (Are you furry? Are you bigger than a . . .? Do you eat acorns?) Once all names are guessed and you are sure that the children are well acquainted with their plant or animal, have them sit down again.

The next part of the activity requires a ball of string. Begin to make food webs. Give one string end to a "plant." The nearest plant consumer is handed the ball of string, which is unwound over to the nearest predator of the plant consumer. Members of the web keep holding onto their position on the string. Continue passing the string until it stops at the top predator. Obviously you may need to make judicious choices for the names of the plants and animals to make the game work smoothly. This game is most effective if there are more plants and primary consumers than top predators, as is the case in nature. Cut the string at the top predator and start the ball again at another plant.

To include all members in a web, some organisms may have to have several consumers, and the top predator may end up holding many strings, but these are accurate reflections of natural principles. Don't let the kids get away with shortcuts or glib answers. Try to be as true to life as possible and encourage controversy (and logical reasoning) in settling differences of opinion. Once all the links are made and all hands have a string, ask the children to raise their hands holding the strings over their heads. An intricate web will be seen.

At this point, while the web is up and the strings are tight, make one of the players "extinct" by cutting the string the child is holding. Perhaps you can make up a realistic reason for the extinction, such as loss of habitat through thoughtless waste disposal. All other members of the food web will be affected. For those that are "dependent" on only that member, the link between it and the predator will also be extinguished, unless there is easy access to another nearby string. Point out that any consumer organism that has been preyed upon by the extinct species may have it easy for a while, but if their numbers go unchecked, they will be more vulnerable to disease from overpopulation. While one loss may do little to damage the web, observe the effect of another or several extinctions. It takes very few losses to damage the original web irrevocably. Give the children a chance to describe incidents or examples they have heard about involving endangered species or habitats. Relate the issues of endangered species to the activity, encouraging the children to verbalize their rationales for protecting habitat integrity and the role of a single species in supporting the others.

Appendix

It is very important that you, as a teacher or parent, be aware that the following summary is not a set description of behaviors of exact age groups. Not only do children go through these stages at different times, but we all tend to repeat the stages as we approach a new body of information. Consider the behaviors as cumulative. Older children will still retain attributes of earlier stages. This is for *general* reference only, as a guide to help you choose activities best suited to your children.

General Age Level	Typical Behaviors and Interests	Appropriate Activities
4 to 6 years	Physical activity is a predominant theme: children are interested in success in new movements, new experiences. They are beginning to learn the benefits of patience to achieve their goals. They are sense-oriented and take in descriptive information very quickly. Their goal is to absorb information, not necessarily to use it in solving problems.	Children want to know "why," mostly in an urge to learn language and vocabulary. Keep your explanations simple. Children enjoy using their senses and attendant vocabulary. They appreciate contrasts and distinctions. They can recognize colors, shapes, textures, and enjoy using the categories. They enjoy collecting, sorting, stacking, making collages.
	Children are interested in concepts related to amounts. They can deal with early math experiences.	They enjoy variety and contrast of activities.
	They are beginning to appreciate seasonal changes.	They are curious about time relationships, but the concept of ongoing time is still flexible and subjective. They enjoy counting and measuring, but don't expect conclusions other than "more" or "less."
	They like to make things happen and are beginning to perceive cause-and-effect relationships. They have excellent photographic memories (which are often lost at puberty).	They enjoy direct action on objects, especially involving physical manipulations. They enjoy matching colors and shapes with found objects.

Ages 6 to 10	Children are increasingly able to work out answers in their head. They also wonder if the things they imagine could really happen. They are eager for more things to imagine.	They enjoy what-if situations and absurdities. This is a time to stretch vocabulary-building and awareness of form and color.
	They are increasingly proud of personal achievements, private stores of information, and special talents. They are willing to stand apart from the peer group as specialists.	They like to investigate alone and report on their findings. They enjoy doing special projects that distinguish them from others.
	They are interested in larger, more abstract categories. They have more concrete experiences for reference and enjoy using them and identifying with them.	They enjoy classification exercises, learning distinctions, and correct nomenclature.
	They are able to refer correctly to seasonal events and patterns.	They are interested in refining their knowledge of what happens when and how life forms change over time. They can appreciate life cycles and food chains.
	They are able to deal with objects and events in a conceptual field of reference.	They enjoy being scientific, taking measurements, and reaching conclusions, but their ability to abstract is still limited to experiences.
	They can relate objects symbolically, although subjectively.	They enjoy amplifying a situation while retaining the basic parameters. They like comparing objects and identifying with other life forms. At the same time, they are increasingly willing to be critical in their evaluations.
	They enjoy playing with crazy comparisons.	
Ages 11 to 17 years	Children's moral and social attitudes emerge and are of great satisfaction. Joined	Objects and identities are seen against a background of "acceptable" and "unac-

Ages 11 to 17 years *continued*	with these is an interest in experimenting with various roles.	ceptable." Children are very much aware of "right answers" and are willing to work to learn if the information has an application in a group context. Helping them hone communication skills is important.
	They can follow and enjoy abstract themes. They are able to think increasingly abstractly. They are able to identify with situations and identities in nature.	They enjoy thinking about and discussing situations from all angles. Encourage role-playing to avoid good/bad connotations. They enjoy testing hypotheses. They enjoy analogies and allusions (also poetry).
	Social pressures demand conformity to group standards.	Children who were exceptional at specializing in information are now reluctant to stand apart from their group. Set up situations where individual contributions enhance group achievements.
	They are aware of their role in environmental issues.	They are willing to take part and support larger issues involving environmental quality. Seek out local situations in which they can take part and contribute as a group.

Glossary

Alternate: Spaced singly along the length of a stem.

Anther: The part of the stamen, usually the tip, that produces pollen.

Biennial: A nonwoody plant that takes two growing seasons to complete its life cycle. It blooms the second season, producing seeds.

Bract: A small, leaflike structure found at the base of a flower cluster.

Carbohydrate: The product of photosynthesis (see chlorophyll). Simple sugars are the initial product. They are sometimes restructured and stored as complex carbohydrates (starches).

Catkin: A conelike flower structure, producing either pollen or seed. They are typical of wind pollinated trees such as pines, willows, and birches.

Chlorophyll: The plant pigment responsible for the transformation of sunlight energy into the storable energy of carbohydrate. The process is known as photosynthesis.

Cleistogamous: The adjective for a flower that never opens and is self-fertilized. Typical of violets.

Compound leaf: A leaf that divides into two or more leaflike parts.

Corolla: The collective term for the showy display of the petals of a flower.

Cotyledon: The area of food storage in a seed.

Cuticle: The semiporous skin of the plant. Usually translucent.

Deciduous: A plant that loses its leaves before the season of drought and/or cold.

Embryo: The baby plant within the seed. Rudimentary root and leaves are usually visible.

Evergreen: A plant that retains its leaves for more than one growing season.

Fiddlehead: The emergent form of any fern frond. Usually coiled in structure.

Fungus: Plants that lack chlorophyll and obtain energy and nutrients chiefly from decaying organic material.

Mushroom: Usually refers to the above ground, spore producing structures of various fungi.

Mycelium: The loosely organized, threadlike structures of the below-ground fungus plant.

Nectar guide: A conspicuous pattern of lines, dots, or contrasting colors that serve to direct a pollinator's movements.

Opposite: Spaced in pairs along the length of a stem.

Osmosis: The movement of water across a cell membrane: from greater concentration of water to a lesser concentration.

Ovary: The base of a pistil, an enclosure protecting the ovules and eventually the fertilized seeds.

Ovules: Female reproductive cells, the unfertilized seeds that lie inside the ovary.

Perennial: A nonwoody plant that dies back to its roots each year. New growth occurs from the roots each growing season.

Photosynthesis: See Chlorophyll.

Pistil: The collective term for the female flower parts. The pistil includes the stigma, the style, and the ovary.

Pollen: The tiny grains produced in the anthers that contain the male reproductive cells.

Pollination; The cross-mixing of genetic material between two plants of the same species. Mixing begins when pollen grains land on the surface of the stigma and ends in the combination of male and female reproductive cells.

Pollinator: An animal vector whose activities serve to transport pollen from one flower to another of the same species in such a way that pollination is achieved.

Prothallus: The small, heart shaped leaf that produces sexual processes and eventually, the spore-making frond of a fern.

Seaweed: A nonflowering plant, an alga, that lives submerged in the ocean.

Sepals: The leaflike structures that enclose the petals and usually form the outer covering for the flower bud. Sometimes sepals are brightly colored and form part of the display.

Simple Leaf: A leaf of only one part.

Speciation: The biological process by which new species are formed as a population responds to the selective pressures of a given environment.

Spores: Single-celled reproductive units typical of fungi, ferns, and mosses. In contrast to seeds, spores contain no food supply for the embryo plant.

Stamen: The collective term for the male reproductive parts. A thin stalk called the filament supports the pollen-producing anther.

Stigma: The expanded tip of the pistil, usually sticky, upon which pollen is deposited.

Style: The stalk connecting the stigma and the ovary.

Venation: The pattern of veins within the tissues of a leaf or other plant structure.

Bibliography

ENVIRONMENTAL EDUCATION

Carin, Arthur A., and Robert B. Sand. *Teaching Science Through Discovery*, 2nd ed. Columbus, OH: Charles E. Merrill Publishing Co., 1970.

Criswell, Susie. *Nature with Art.* Englewood Cliffs, NJ: Prentice-Hall, 1986.

Great Britain Schools Council. *Early Experiences. The Nuffield Science Series.* London: MacDonald Educational, 1972.

———. Trees 1 and 2. *The Nuffield Science Series.* London: MacDonald Educational, 1972.

Gross, Phyllis, and Esther P. Railton. *Teaching Science in an Outdoor Environment.* Berkeley: University of California Press, 1972.

Jenkins, Peggy Davison. *Art for the Fun of It: A Guide for Teaching Young Children.* Englewood Cliffs, NJ: Prentice-Hall, 1980.

Levenson, Elaine, *Teaching Children About Science.* Englewood Cliffs, NJ: Prentice-Hall, 1985.

Link, Michael. *Outdoor Education: A Manual of Teaching in Nature's Classroom.* Englewood Cliffs, NJ: Prentice-Hall, 1981.

Nicklesberg, Janet. *Nature Activities for Early Childhood.* Reading, MA: Addison Wesley, 1976.

Rockwell, Robert E. *Hug a Tree: And Other Things to Do Outdoors with Young Children.* Mt Rainier, MD: Gryphon House, 1985.

———. *Mudpies to Magnets: A Preschool Science Curriculum.* Mount Rainier, MD: Gryphon House, 1987.

Roth, Charles E. *The Plant Observer's Guidebook: A Field Botany Manual for the Amateur Naturalist.* Englewood Cliffs, NJ: Prentice-Hall, 1984.

Russell, Helen Ross. *Ten-Minute Field Trips: Using the School Grounds for Environmental Studies.* Chicago: J.G. Ferguson, 1973.

Scott, Jane. *Botany in the Field.* Englewood Cliffs, NJ: Prentice-Hall, 1984.

Sisson, Edith A. *Nature with Children of All Ages: Activities and Adventures for Exploring, Learning, and Enjoying the World Around Us.* Englewood Cliffs, NJ: Prentice-Hall, 1982.

Skelsey, Alice and Gloria Huckaby. *Growing Up Green.* New York: Workman Publishing Company, 1973.

Suzuki, David. *Looking at Plants.* New York: Warner Communications Company, 1985.

Swan, Malcolm. *Tips and Tricks in Outdoor Education*, 2nd ed. Danville, IL: Interstate, 1979.

University of California. 1970–80. OBIS: *Outdoor Biology Instructional Strategies.* Nashua, NH: Delta Education.

van Matre, Steve. *Acclimatization.* American Camping Association. Martinsville, IN: Bradford Woods, 1972.

Wensburg, Katherine. *Experiences with Living Things.* Boston: Beacon Press, 1966.

FLOWERS

Written for children and recommended for adults

Busch, Phyllis S. *Lions in the Grass: The Story of the Dandelion, a Green Plant.* New York: World Publishing, 1968.

Heller, Ruth. *The Reason for a Flower.* New York: Grosset & Dunlap, 1983.

Hutchins, Ross E. *This Is a Flower.* New York: Dodd, Mead, 1963.

Overbeck, Cynthia. *Sunflowers.* Minneapolis: Lerner Publications, 1981.

McMillan, Bruce. *Apples, How They Grow.* Boston: Houghton Mifflin, 1979.

Milne, Lorus and Margery. *Because of a Flower.* New York: Atheneum, 1975.

Rahn, Joan Elma. *How Plants Are Pollinated.* New York: Atheneum, 1975.

Selsam, Millicent E., and Jerome Wexler. *The Amazing Dandelion.* New York: William Morrow, 1977.

Written for adults

Barth, Friedrich G. *Insects and Flowers: The Biology of a Partnership.* Princeton, NJ: Princeton University Press, 1985.

Dowden, Anne Ophelia. *The Secret Life of Flowers.* New York: Odyssey Press, 1964.

Gibbons, Bob. *How Flowers Work: A Guide to Plant Biology.* New York: Sterling Publishing Company, 1984.

Knobel, Edward. *Field Guide to the Grasses, Sedges and Rushes of the United States.* New York: Dover Publications, 1980.

Meeuse, Bastiaan, and Sean Morris. *The Sex Life of Flowers.* New York: Facts on File, 1984.

Platt, Rutherford. *Our Flowering World.* New York: Dodd, Mead, 1947.

SEEDS

Written for children and recommended for adults

Budlong, Ware. *Experimenting with Seeds and Plants.* New York: G.P. Putnam's Sons, 1970.

Hutchins, Ross E. *The Amazing Seeds.* New York: Dodd, Mead, 1965.

Johnson, Hannah L. *From Seed to Jack-O'Lantern.* New York: Lothrop, Lee & Shepard, 1974.

Petie, Harris. *The Seed the Squirrel Dropped.* Englewood Cliffs, NJ: Prentice-Hall, 1976.

Rhan, Joan Elma. *How Plants Travel.* New York: Atheneum, 1973.

Selsam, Millicent E. *Play with Seeds.* New York: William Morrow & Co., 1957.

Written for adults

Phillips, Harry R. *Growing and Propagating Wildflowers.* Chapel Hill, NC: University of North Carolina Press, 1985.

FUNCTIONS OF PLANT PARTS AND BASIC GROWTH FACTORS

Written for children and recommended for adults

Hutchins, Ross E. *This Is a Leaf.* New York: Dodd, Mead, 1962.

Kurtz, Edwin B., and Chris Allen. *Adventures in Living Plants.* Tucson: University of Arizona Press, 1965.

Nussbaum, Hedda. *Plants Do Amazing Things.* New York: Random House, 1977.

Rahn, Joan Elma. *More About What Plants Do.* New York: Atheneum, 1975.

Selsam, Millicent E. *Play with Plants.* New York: William Morrow & Co., 1956.

Stone, A. Harris. *Plants Are Like That.* Englewood Cliffs, NJ: Prentice-Hall, 1968.

Written for adults

Platt, Rutherford. *This Green World.* New York: Dodd, Mead, 1942.

Rickett, Harold W. *Botany for Gardeners.* New York: Macmillan Co., 1971.

Stone, Doris M. *The Lives of Plants.* New York: Charles Scribner's Sons, 1983.

Vallin, Jean. *The Plant World.* New York: Sterling Publishing Co., 1967.

Wilson, Rom. *How Plants Grow.* New York: Larousse and Co., 1980.

PLANTS CHANGE IN RESPONSE TO SEASONAL CHANGES AND EVENTS

Written for children and recommended for adults

Selsam, Millicent E. *Maple Tree.* New York: William Morrow & Co., 1968.

Spier, Peter. *Rain.* New York: Doubleday & Co., 1982.

Written for adults

Brown, Lauren. *Weeds in Winter.* New York: Norton, 1976.

Russell, Helen Ross. *Winter Search Party.* New York: Nelson, 1971.

Stokes, Donald W. and Lillian. *A Guide to Enjoying Wildflowers.* Boston: Little Brown, 1985.

PLANTS USED BY PEOPLE

Written for children and recommended for adults

Gutnik, Martin J. *How Plants Make Food.* Chicago: Children's Press, 1976.

Rahn, Joan Elma. *Grocery Store Botany.* New York: Atheneum, 1978.

Selsam, Millicent E. *The Plants We Eat.* New York: William Morrow & Co., 1981.

Written for adults

Anderson, Edgar. *Plants, Man and Life.* Berkeley: University of California Press, 1967.

Phillips, Roger. *Wild Food.* Boston: Little, Brown, 1986.

Sloane, Eric. *A Reverence for Wood.* New York: Ballantine Books, 1965.

THE PLANT IN RELATION TO ITS ENVIRONMENT

Written for children and recommended for adults

Gallab, Edward. *City Leaves, City Trees.* New York: Charles Scribner's Sons, 1979.

Harrison, Hal H. *A Field Guide to Birds' Nests (found East of the Mississippi River).* Peterson Field Guide Series. Boston: Houghton Mifflin, 1975.

___. *A Field Guide to Western Birds' Nests.* Peterson Field Guide Series. Boston: Houghton Mifflin, 1979.

Hutchins, Ross. E. *Galls and Gall Insects.* New York: Dodd, Mead, 1969.

Mabey, Richard. *Oak and Company.* New York: William Morrow & Co., 1983.

Selsam, Millicent E. *Plants that Move.* New York: William Morrow & Co., 1962

Written for adults

Watts, May Theilgaard. *Reading the Landscape of America.* New York: Collier Books, 1975.

Index

Acorns, 70–71
Age of trees, computation of, 105–6, 109
Amaryllis, 57–58
Animals, as seed carriers, 66
Anthers of flower, 43, 50, 57
Aquatic plants, 87
Auxins, 91

Bacteria, nitrogen-fixing bacteria, 63, 94
Beans
 cooked/uncooked, comparison of, 78–79
 micropyle, 75
Bees
 bee-pollinated flowers, 48
 distributing/receiving pollen, 50–51
 food of young bees, 49
 stinging, 49
Berries, 67–68
Biennials, 137
Birds, berry-eating, 68–69
Body language in plants, 126–27
 plant movement, 126–27
Body language of student, 5
Bract of flower, 44
Buds
 setting, period for, 104
 terminal and auxiliary, 91
Burdock, 66
Burrs, 66

Carbohydrates, 76
Carbon dioxide, 92
Catkins, 53
Centering, approaches to, 24–25
Chloroplasts, 115
Cleistogamy, 52
Collections, collecting methods, 26–28
Common Blue Violet, 52
Corn, pollination, 55
Corolla of flower, 44
Cotyledons, 76
Cranberries, 67

Desert plants, 87, 88
Devil's stick-tight, 66
Disruptive students, 5

Environmental awareness
 environmental values, development,
  28–29
 learner-centered approach, 11
 literature and, 29

plants activities and, 3–4
 teaching philosophy, 5
Ethylene gas, 74

Feely bag, use of, 10
Ferns
 fiddlehead ferns, 38
 propagation of, 38
Fiddlehead ferns, 38
Floating seeds, 66–67
Flowers, parts of, 43–45
Fly-pollinated flowers, 48
Frames (cardboard), use of, 10
Fruits compared to vegetables, 60–61

Garden in the Woods, 4
Gnat-pollinated flowers, 48
Grasses
 growth patterns, 54
 pollination, 55

Hormone powders, root growth, 117
Hummingbird-pollinated flowers, 48

Iodine, for starch testing, 77, 79

Learners
 age level/typical behaviors/activities,
  145–47
 etiquette and, 6
 individual styles of, 10–11
 self-directed discovery and, 6
 senses, use of, 6, 7
 teacher as model to, 7
 as teachers, 25–26
Leaves, yellowing, 113

Micropyle, 75
Mistletoe berries, 68
Mushrooms
 cautions about, 39
 identifying species, 39–40
Mycelium, 39–40
Mycologists, 39

Names of plants, use of, 29
Native Americans, 25
Nectar guides of flowers, pollination,
 47–48
Nitrogen-fixing bacteria, 63, 94

Osmometer, 95
Osmosis, 94–95

Ovules of flower, 43, 56–57
Oxygen, 92–93

Photosynthesis, 92
Pinecones, 53
Pistil of flower, 43
Plant food, 58, 113
Pollination
 corn, 55
 flower's mechanisms for
  distributing/receiving pollen, 50–51
 grasses, 55
 nectar guides of flowers, 47–48
 pollinators and flower characteristics,
  48, 49
 self-pollinated flowers, 52–53, 57–58
 weather conditions and, 51
 wind-pollinated flowers, 53–54
Pollution, plants as indicators of, 128–29
Pots for plants
 size factors, 111–12
 starting seeds, 112
Pussy willows, 53

Quadrat, 131–32

Random samples, 132
Roots, healthy vs. dead, 113
Row plants, as environmental indicators,
 127–28

Seeds
 berries, 67–68
 floating seeds, 66–67
 hitchhiking seeds, 65–66
 ripeness of, 63
 starting seeds, 112
 traveling seeds, types of, 20
 windborne seeds, 64–65
 See also Beans.
Self-pollinated flowers, 52–53, 57–58
Senses
 enhancing observational skills, 22–23
 use of, 6, 7
Sepals of flower, 44
Shrubs, branching patterns, 33
Soil, potting soil, 113
"Solar panel" surface of tree, 101–2
Special needs children, adaptations for,
 7–8
Spread sheet, use of, 10
Sprouting, observation of, 77–78
Stamens of flower, 43, 50, 57

Starch
  beans, testing for, 77
  in plants, 78–79
Stigma of flower, 43, 50–51, 53, 55, 57
Stomata, 85
Style of flower, 43
Symbiotic relationships, beans and
  bacteria, 63

Tannin, 71
Taproots, 137
Teachers, as model to learners, 7
Tick-Trefoil, 66
Transect, 131–32
Trees
  branching patterns, 33
  ethylene gas used as defense, 74

Vegetables compared to fruits, 60–61

Weather, effect on pollination, 51
Windborne seeds, 64–65
Wind-pollinated flowers, 53–5

Xylem, 108

Activities Index

Acorns, 70–7
  collecting/sorting, 71
  dying activity, 71
  whistles from, 71
Acting, plant growth activity, 19
Activities
  excursions, structuring of, 8–9
  learning enhancing devices, 10
Amaryllis, growing from seed, 57–58
Animals, imaginary, creation of, 138–39
Art activities
  general types of, 12–15
  leaves, 88–91
  soil, 120–21
  trees, 100–101
  value of, 11
  See also specific types of art activities.
Autumn leaf activities, 102–3

Beans
  sprouting beans, 75–76
  starch testing, 77
Berries, attracting berry-eating birds,
  67–68
Birds
  attracting berry-eating birds, 67–68
  bird nests, examination of, 142–43
Book-making, about rain, 137
Buds
  forcing open, 105
  investigation of, 98–99
Bulbs
  onion bulbs, 100
  paperwhite narcissus, 99–100
Calendar making, calendar of bloom, 47
Chart/graph activities, 129–30
Collections
  collecting bag, use of, 10
  scavenger hunts, 27–28
  seeds, activities with, 59–60
Color samples, use of, 10

Comparisons, sensory activities and, 18
Cork, investigation of, 107
Cranberries
  drink-making, 67
  separating good berries from rejects, 67
Creature making, 14

Dandelions, study of, 135–36
Data sampling methods, 132–33
Dead wood
  beetle tunnels in, 141–42
  rotting log, investigation of, 140–41
Drawing, 12
Dyeing activities
  dye bath, preparation of, 16–17
  with plant materials, 15–17
  sources of colors, 17
  tannin from acorns, 71

Ecosystems, creating two worlds, 19–20
Edible activities
  bean sprouts, 75–76
  cranberry juice, 67
  pumpkin seeds, 74
  raisins, 72–73
Sumac-ade, 69–70

Fern nursery, 38
Fertilizers and plants, experiments,
  122–23
Flowers
  dandelions, study of, 135–36
  garden flowers compared to wildflowers,
    46
  inventing flowers, 46–47
  of grasses, observation of, 54–55
  miniatures, collecting, 42–43
  observing seeds of, 56–57
  picking, consequences of, 41
  pressed flowers, 14
  sorting, 45

superflowers, creation of, 48–49
  winter rosettes, search for, 136–37
  See also Pollination.
Fruits
  ripening with ethylene gas, 74
  comparing to vegetables, 60–61

Gases and plants
  gas release, 86–87
  oxygen and growth activity, 93
Grapes, making raisins, 72–73
Grasses, flowers of, observation, 54–55
Growing plants
  fertilizers, 122–23
  light response experiments, 114–15
  no growth, common reasons for, 113–14
  pots, 111–12
  potting soil, 113
  propagation from cuttings, 116–17
  soil activities, 117–22
  starting seeds, 112
  terrariums, 123–24
  variables influencing growth, 125
Growth curves and averaging data,
  130–31

Habitats
  awareness activity, 133
  habitat hats, 13
Herbs, growing from seeds, 60

Identification activities
  game for developing plant-related
    vocabulary, 31–32
  key for species identification, 32–33
  parent/offspring plant search, 68–69
  parts of flowers, 43–45
  tree identification games, 30–31
Insects
  beetle tunnels in dead wood, 141–42
  damage from, observation of, 139–40

Key, for species identification, 32–33

Leaves
  autumn leaf activities, 102–3
  colored smear activities, 89–90
  evergreen leaves compared to deciduous leaves, 110–11
  gas release, 86–87
  preservation of, 84, 90
  pressed, activities for, 90–91
  print making, 88
  silhouettes, 89
  skeletons, preparation of, 84–85
  transpiration, investigations of, 85–86
  water storage activities, 87–88
Light and plants
  light as growth variable, 125
  light response experiments, 114–15

Magnifying lenses, use of, 6, 10, 18
Masks, 14
Materials, learning enhancing devices, 10
Metaphors, nature metaphors, 23
Mobiles, 12
Model making, 13
  mushrooms, 40
Mosaics, 12
Mosses, haircap moss rain meter, 40–41
Mushrooms
  model making, 40
  spore prints, 39–40

Nature walk
  question walk, 19
  small worlds, exploring, 18–19

Observation, diary of, 134
Onion bulbs, 100
Osmosis, measurement of, 94–95

Painting, 12
Paperwhite narcissus, growing from bulbs, 99–100
Peanuts, growing from seeds, 62–63
Pinecones
  locating seeds, 72
  rain meter from, 72
Plant activities
  buds, 98–99
  bulbs, 99–100
  chart/graph activities, 129–30
  data sampling methods, 132–33
  diary of observation, 134
  growth curves and averaging data, 130–31

imaginary plants, creation of, 138–39
insect damage, observation of, 139–40
leaves, 83–90
plant parts game, 100
plants as historical indicators, 138
plant succession, observation of, 131–33
roots, 92–97
row plant responses, 128–29
seasonal change activities, 133–34
stems, 91–92
veins, 83
See also Growing plants; activities listed under specific part of plant.
Pollination
  action, imitating, 50
  corn, observation, 55–56
  flowers to attract bees, designing, 49
  self-pollinated flowers, observing, 52–53
  self-pollinating an amaryllis, 57–58
  weather conditions, observing, 51
  wind-pollinated flowers, observing, 53–54
Predators, game related to, 143–44
Pressed leaves and flowers, 14, 90–91
Prints, 13–14
  leaf prints, 88
  mushroom spore prints, 39–40
  seaweed prints, 37
Propagation from cuttings, 116–17
Puppets, 14

Rain, book-making about, 137
Rain meter
  from haircap moss, 40–41
  from pinecones, 72
Raisins, preparation from grapes, 72–73
Roots
  length, calculation of, 97
  power, testing of, 93–94, 95–97
Root vegetables, observation of, 97–98
Rubbings, 13
  beetle tunnels in dead wood, 141–42

Scavenger hunts, 27–28
Seasonal change activities, 133–34
Seeds
  acorn activities, 70–71
  amaryllis grown from, 57–58
  collections, activities with, 59–60
  floating seed activities, 66–67
  of flowers, observing, 56–57
  herbs grown from, 60
  hitchhiking seed activities, 65–66
  kitchen sources, sprouting seeds, 58–59
  peanuts grown from, 62–63

pinecones, locating seeds, 72
pumpkin seed observation, 72–73
ripeness, observation of, 63–64
seed coat, observing, 74–75
Sumac-ade, preparation of, 69–70
traveling seeds, 20–21
vegetables compared to fruits, 60–61
windborne seed activities, 64–65
Sensory activities
  comparison and, 18
  general types of, 18–20, 21–22
  soil-related, 117–18
Silhouettes, leaves, 89
Soil
  acid/base testing, 121–22
  art activities, 120–21
  exploratory activities, 117–22
Stems of plants, damage, observing response to, 91–92
Sumac, drink-making, 69–70

Terrariums, 123–24
Texture clues, use of, 10
Thermometer, use of, 10
Trees
  age, computation of, 105–6, 109
  art activities, 100–101
  autumn leaf activities, 102–3
  buds, forcing open, 105
  cork, investigation of, 10
  evergreen leaves compared to deciduous leaves, 110–11
  height estimation, 103–4
  leaves, computing number of, 101–2
  wood, investigation of, 108–10

Vegetables
  comparing to fruit, 60–61
  root vegetables, observation of, 97–98
Veins of plant, water movement activity, 83–84

Water and plants
  movement through veins, 83–84
  osmosis, measurement of, 94–95
  storage of water, 87–88
  transpiration, investigation of, 85–86
Weavings, 12
Whistles, making from acorns, 71
Winter rosettes, search for, 136–37
Wood
  investigation of, 108–10
  See also Dead wood.

About the Author

Jorie is an Easterner who grew up exploring the land and plants of Georgia, Virginia, Connecticut, and Maine. She continued her explorations of plants during several journeys to the West and while attending college in the Midwest. Since she was a teenager, she has been working with children in the out-of-doors and has learned much from their wonder and their questions. Her main teacher at this time is her three-year-old son, Tyler Ray.